IMPOSSIBLE MISSION

IMPOSSIBLE MISSION

2nd Edition

Kurth Krause
2023

Second Edition: June 2023

This novel is a work of fiction. The characters, dialog, and events in this book are the products of the author's imagination. The description of the known Universe and the laws of physics are, however, factual. With one exception, any similarity to actual persons living or dead is coincidental and are not intended by the author.

Cover design by Taylor Odish

ISBN 978-0-4568-6-7

Contents

Preface

When I began writing *Impossible Mission,* I intended it to be both entertaining and educational. But the plot naturally called out for more. Discussions of a political nature screamed to be introduced by each member of the crew. To my surprise, this need carried throughout the book. It was also natural to introduce religion into the observations of our fantastic Solar System and this amazing Universe. Most surprising to me was the degree of philosophical thought which ultimately permeated throughout the theme of the book and, in fact, became a theme in itself.

I chose to forgo the typical units of science in describing distance, speed, mass, and force, instead using feet, miles, miles-per-hour, and pounds to better enable the American reader to relate to more familiar units of measure. I did my best to call on my background in astronautical engineering and orbital mechanics to describe the Solar System, the extraterrestrial phenomena, and the experiences with the laws of physics to be consistent with reality. While I took liberty extrapolating events that have not yet happened, the storyline is within the realm of the possible, which we may someday experience.

I hope my readers are able to immerse themselves in the impossible mission adventure as much as I did when writing it.

1. A Calling

Commander Kevin Owens waited patiently for the countdown. The Launch Control Center had announced a hold. He lay face up in the SpaceX Dragon couch molded perfectly to his firm, trim, muscular body. He was proud to wear the new streamlined modular spacesuit with his original embroidered mission patch on one side and the NASA logo patch on the other. He was about to begin a mission from which he would likely never return. The media declared it an impossible mission. But instead of being apprehensive, he was excited, maybe euphoric. *My extraordinary crew and I will prove them wrong,* he vowed.

As he proudly looked over at his personally selected crew, he wondered whether they, like he, were reflecting on their personal journeys that brought them to this point.

***

For Kevin, it really started eleven years earlier with the conversation with his dad after he and his parents returned home to Southern California after a two-week summer vacation touring the East Coast. Kevin was fifteen and had just completed his junior year in high school.

"Mom, Dad, I've decided where I want to go to college."

"That's great Kevin," responded his dad, enthusiastically. "Harvard? MIT?"

"Nope. I've decided to apply to the Military Academy at West Point."

"What?" exploded his dad. "You want to use your 151 IQ to become a soldier? Kevin, you can get into any school in the country! I hope you weren't blinded by the glamour and prestige you saw at the West Point displays."

"Dad, we can't afford those schools. They cost as much as $300,000 for an undergraduate degree."

"I know they do, but we have saved over $39,000 for your education and will save more the next five years. You may be able to get a partial scholarship with your top grades, Scholastic Aptitude Test results, and extracurricular activities. Our income is not high enough to disqualify you. Plus, you can apply for a student loan, which we will help pay. We don't want you to limit yourself with a military career."

"But my career would not be limited at all. America's top generals throughout history, like Norman Schwarzkopf, Douglas MacArthur, George Patton and Robert E. Lee were West Point Graduates. So were

President Eisenhower and Ulysses Grant. Apollo astronauts Buzz Aldrin, Michael Collins, Ed White, and Frank Borman all graduated from West Point. So did Secretary of State Mike Pompeo.

"Dad, making a lot of money is not my top priority. I want to do what I can to give back to my country like these famous patriots did."

This resulted in a long pause. Kevin's dad did not realize a fifteen-year-old could develop such maturity. And be so altruistic. He was embarrassed for not recognizing this, but also so proud of his son.

"Okay, Kev," concluded his dad. "But let's apply to Harvard, MIT, Stanford, and Caltech just in case you don't get accepted at West Point."

"Deal," responded Kevin, beaming.

The next day, Kevin called the Army recruiting office to arrange a meeting with the academy liaison officer.

"Kevin, your qualifications are impeccable. We weigh grades at 60%, high school athletics and leadership 30%, and citizenship 10%. Your grades are outstanding and your choice of courses more than satisfy our requirements. As captain of the golf team and drum major for the high school band, you certainly have demonstrated leadership. We will check out these and the remainder of the qualifications by interviewing your teachers. You need to fill out this application with all this information. Also, you need to be nominated by a member of Congress, but that should not be a problem because each congressperson is allowed ten nominations. You can submit this same application form to your congressperson because he uses the same criteria as the academy. So, the only issue is your age. You must be at least seventeen unless you receive a waiver. We can apply for it.

"But you should know that there are more than ten qualified applicants for each opening, so it won't be easy to secure an appointment. Also, I don't know how much you know about the rigors required of each cadet. It can be tough. More than 20% do not make it through their first year. You must complete the training in the four years allocated. Training consists of four pillars of excellence, each worth 25%: military training, academics, athletics, and character development, which includes religion, patriotism, and camaraderie. This requires a full workload of 12 hours each day.

"If you receive a nomination and accept, you must complete all the requirements in four years. If you drop out during the first two years, there is no consequence. But if you drop out after beginning your third year, you are on the hook for repayment. You can repay the hundreds of thousands of dollars invested in you by the Army or repay with military service. But you will enter the Army as a private for your military service repayment. Virtually every cadet has considered dropping out at some point because the experience can be grueling. That is why you are not granted home leave for the first four months because you may decide not to come back.

"The rewards, however, are commensurate with the challenges. The government pays for all your tuition, room, and board for all four years. In addition, you receive a small monthly stipend with which you pay for your uniform, dry cleaning, and incidentals. The value of all this is over $400,000. Yes, it is significantly more than four years at the top schools in the country. Your education goes well beyond what you would receive at any university. You graduate commissioned as a second lieutenant in the United States Army and serve a minimum of five years."

Now, I know this is the right choice, thought Kevin. He was impressed with everything he heard.

After the liaison officer left, Kevin's dad said, "Let's consider the Air Force Academy in Colorado Springs too. I think you'd prefer fighting a war in the cockpit of an airplane, rather than a foxhole."

"Okay, Dad, that makes sense to me."

The Air Force liaison officer laid out the almost identical requirements, but added, "Upon graduation, you have a choice of the Air Force or the Space Force. The latter also provides a pathway to the astronaut corps. After graduation you have a five-year commitment to service as an Air Force or Space Force officer. But if you complete pilot training, your minimum commitment will be ten years."

The astronaut path caught Kevin's attention, so he applied to both academies immediately. He also applied for and received both academy nominations from the House Member from his congressional district. One year later he received an early appointment from both academies, having been granted a waiver for his age since he would be only sixteen at the time he begins.

"Dad, I just saved us $300,000. I turned down the offers from the universities and intend to accept the appointment to the Air Force Academy. I barely made the minimal weight requirement of 129 pounds, thanks to Mom's great milk shakes. They have invited all early appointees to an orientation visit next month."

"I hope you are making the right decision, Kevin. But we support you completely. This is a big challenge. If it doesn't work out, I guess you could apply to the universities again later."

***

Kevin reveled in the camaraderie of attending two days of introductory orientation at the academy's

Colorado Springs campus before final acceptance of the appointment. This enjoyment was amplified in his six weeks of "boot camp," designed to orient first-year cadets to the tough academy training wherein some cadets muster out.

Kevin was overjoyed when he made the academy golf team as a cadet first-year student (a Doolie), partly because he earned the privilege to have his meals with the team, avoiding the harassment at every meal imposed by the upper classmen in his squadron. His varsity golf more than fulfilled the athletics pillar required by the academy.

The first two years he, like all Cadets, took the required core curriculum of 50% math and science, and 50% "fuzzy" courses, such as history, English, political science, economics, leadership, ethics, law, behavioral science, and philosophy. He satisfied his foreign language requirement with Mandarin, as suggested by his counselor.

He also was required to round out the other two pillars of cadet training: miliary training and character building. At 6:00 A.M. each morning he attended services at the academy's beautiful, famous Cadet Chapel.

Kevin's most challenging and memorable training was Survival, Evasion, Resistance, and Escape (SERE). Whether it is in the jungle, the desert, the arctic, at sea, or as a prisoner of war, each cadet must be prepared to survive, evade, resist, and escape any situation. They had to be ready for anything and must return with honor. He and a fellow cadet were dropped into the wilderness from a helicopter with no modern means of sustaining life. The Survival Segment required that he sustain himself with whatever plants, wild animals (e.g., rabbits), and water he could find in the wild. The Evasion Segment was to evade the enemy who were searching to capture him. Ultimately upon being captured, the Resistance Segment required that he divulge only his name, rank, and serial number no matter what. He was thrown into a stuffy box, with no amenities, which was so small it forced him to stand hunched over, then try to sit on a two-by-four balanced crosswise on an upright two-by-four. He was sleep deprived and would fall over each time he began to fall asleep. During his period of captivity, he was required to attempt to escape by any means possible. Kevin was exhausted when he completed SERE training but emerged stronger, confident, and more resilient than he thought possible.

It was during his junior year that he set his goal to become an astronaut. Ever since President Trump created the US Space Force as a separate branch of the military, it was no longer required to become a test pilot prior to astronaut training. He eagerly chose his major in astronautical engineering, duplicating his father's major at UCLA. There he mastered the difficult fields of orbital mechanics, celestial navigation, propulsions systems, flight control, rendezvous, station-keeping, payload deployment, life support systems, and state-of-the-art flight computers. He was particularly fascinated with astronomy, learning about the Solar System, the peculiarities of the planets and their quirky moons. *If only I could visit them to see their wonders close-up,* he dreamed. He wondered about the stars in the Milky Way and their planets. He wondered if any of them hosted some sort of life.

When he returned home for the first of his allotted short visits and told his father about his curriculum, his father beamed.

"I had no idea the academy gave you such an outstanding education. At UCLA I did not have time for much beyond my math and science classes. The academy expects so much more from you. I wish I had learned more about economics, leadership, communication, and behavioral science. I had to learn these in management training programs in graduate school. I am so proud of you Kev; you certainly made the right decision as far as your education is concerned. I can't wait to learn about the coursework in your major. What an opportunity. And you still have time for intercollegiate golf, military training, and all the rest. How do you do it all?"

"Do you get enough sleep?" questioned his mother.

"Sure. I get at least seven hours sleep every night, so I am awake in class and can absorb the interesting lectures."

"We are so proud of you," echoed his parents.

***

Owens excelled in every course. His almost perfect memory allowed him to listen to the lectures without taking any notes. Unlike the other cadets, he could study sufficiently until "lights out" at 10 pm, while his roommates studied after hours, depriving them of sleep.

But not all his experiences were winners.

All cadets were required to solo in a glider their senior year. After nine flights with his instructor in the cockpit, it was time for his solo flight. Kevin's glider, the TG-16A, was towed to 3,000 feet and released. He was aloft for almost two hours, being lifted by thermals in the warm Colorado Springs atmosphere. He was ecstatic as he soared upward to 6,000 feet, feeling free as a bird. *This must be how it feels in freefall in Space,* he thought. But suddenly his glider hit a downdraft and he lost control. The glider pitched nose down and began diving rapidly. At first Owens did not panic, but after dropping more than 5,000 feet, he thought he was going to die. His instructors watched in horror as the glider continued dropping precipitously, diving downward. Finally, the air pocket ended, and he regained control. He tried to land at the designated point, but he was too low and missed it badly. When the instructors came running to his plane, he begged for another chance. "I'm sorry. Please let me try again."

"Owens, you idiot, the skill you demonstrated saved your life."

Kevin made lifelong friends in his squadron. He told his dad, "I would lay down my life for any of these guys."

Less than a year later his parents were thrilled to attend his graduation ceremony. They proudly pinned the second lieutenant bars to his shoulders, listened to his valedictorian speech, and intently watched him receive his diploma with the other 96 Distinguished Graduates directly from the President of the United States. The entire class of 970 new graduates were enthralled to watch the air show provided by the six F-16 Thunderbirds, the Fighting Falcons, perform over Falcon Stadium. The new graduates threw their hats up during the Thunderbirds' first flyover. But the best was the three flyovers of the SR-71 Mach 3+ Blackbird. After its third pass, it flew vertically straight up into clouds and out of sight. The graduation crowd also loved the performance of the Air Force mascot, the live falcon, in the middle of the field.

Although Kevin chose the US Space Force over the Air Force, he still was required to successfully complete one year of pilot training, recognizing this extended his military commitment to ten years. He first learned to fly the T-6 Texan single-engine turboprop for four months, followed by eight months in the T-38 Talon, (also called the White Rocket). The T-38 is a hypersonic, twin-engine jet requiring a 200-knot landing speed to avoid stalling due to its small wings. This was a blessing for Owens because he knew it was the aircraft used by the astronauts to fly between their simulators and trainers in

Texas and Florida. Kevin told his parents, "I love this plane. I can't believe they're paying me over $3,000 per month to fly it. I would do it for free if I didn't have to eat."

Throughout his year of pilot training, he made up for lost time on the dating scene. At age 20, he was not ready to settle down. The pretty female groupies were happily competing for his attention. His handsome features, dark close-cropped hair, trim six-foot frame and flight suit made him a chick magnet. And the hot Tesla he leased didn't hurt. He selected the Model 3, which gave him the desired performance, but opted to forego the optional $15,000 self-driving hardware because he loved to personally drive the car. He was particularly attracted to a sexy redhead but avoided getting tied to one girl. He was having too much fun.

Owens once again graduated at the top of his class as a pilot. He proudly pinned on his wings. But although the Space Force pilot training prowess enabled him to choose any fighter plane, he announced he wanted to become an astronaut instead and applied to NASA. Yet, despite his accomplishments at an early age, he somehow remained humble. He made friends easily.

At age 21 he was accepted as the youngest in the astronaut program. NASA sent him to MIT for his master's degree in astronautical guidance. He quickly learned the new advances in inertial guidance, orbital mechanics, and numerical integration. He was fascinated by the concept of a spacecraft achieving additional speed without thrust via a gravity assist: swinging by the planets on their trailing side and getting a gravitational boost in speed from their momentum. He wrote a computer program using MIT's HAL language to simulate a grand tour mission of the outer planets using gravity-assist boosts exerted on the free-flying spacecraft. He also learned about a new propulsion system using a nuclear-powered engine to provide continuous spacecraft acceleration.

He continued to enjoy the charms of the pretty college girls in Cambridge. His status as an astronaut enhanced his attraction by the softer sex.

But his adventures at MIT were interrupted by a tragedy at home. Kevin's parents, whom he deeply loved, were killed in a light airplane crash over the San Gabriel Mountains as they were returning from a vacation. This left Kevin virtually alone, as he had no siblings. He was devastated. He prayed for Jesus to fill the hole in his heart. He dove into his work at school to escape the longing created by the tragedy.

That work ethic continued when he returned to the Johnson Space Center to begin his astronaut training. He had rapidly progressed in the Space Force hierarchy with promotions to first lieutenant, captain, and major well "below the zone" (the minimal duration expected for the next promotion). His rapid military promotions resembled the honors bestowed on the first human in space, Cosmonaut Yuri Gagarin. Nikita Khrushchev awarded Gagarin a double promotion from lieutenant to major when he orbited Earth in the Vostok spacecraft on April 12, 1961

Owens rented an apartment in El Lago just four miles east of the Johnson Space Center campus. Tiny El Lago, with 3,000 inhabitants, had been the city of choice for most of the astronauts over the past 50 years.

His classroom training was at JSC as well as Cape Canaveral, where he learned the intricacies of the primary simulator for the SpaceX Dragon capsule. Once again, Owens loved flying his beloved T-38 between Houston and the Cape.

NASA had scrapped their plans to use the Space Launch System and its Orion capsule with the SpaceX booster and its Dragon capsule because the latter had proven its reliability while SLS and Orion had slipped its schedule many times. Its costs escalated dramatically. An Artemis launch using the SLS rocket cost $4.1 billion, while an Artemis launch with SpaceX was just $10 million. The rapid success of SpaceX, a private enterprise founded by Elon Musk, accelerated past the SLS Program with unheard of speed.

The astronauts used the Reduced Gravity Walking Simulator to learn mobility on the Moon. They also spent time in the reduced-gravity aircraft to experience periods of near zero-G environment for short, 20-second durations. This was dubbed the "Vomit Comet" due to the airsickness experienced by most passengers. But Kevin was never sick. He loved the sensation of weightlessness during the 20-second parabolic arcs. Classroom training included learning about Starship's Raptor chemical propulsion engines, its thermal control system, its power generation and distribution system, and Dragon's communication and life support systems. In addition, astronauts studied orbital mechanics, astronomy, rendezvous, and spacecraft docking. Much of this was a review for Owens from his master's degree program at MIT.

NASA had a special mission in mind for Owens, which was not disclosed. NASA selected him to be part of the third staffed Artemis mission to the colony on the Moon, but he would only stay one month, instead of the indefinite durations of the other colonists. NASA was currently colonizing the Moon with fourteen people at the Artemis Base Camp, living in two large, manufactured igloo-like habitats near the Shackleton Crater at the Southern Pole. These colonists had diversified ethnic backgrounds as well as different workload specialties.

He was overjoyed with his new assignment but was quite curious as to why it would be so temporary and what NASA had in mind for his future. He tried unsuccessfully to find out, but, as a good officer, accepted their orders without question.

NASA's SOPHIA observatory had located molecular water on the Sunlit lunar surface to go with the abundant water ice near the poles. This made the proposal of a lunar colony feasible. Building a processing plant on the lunar surface not only produced the water necessary for human and plant consumption but enabled the separation of water into its two components: oxygen and hydrogen. The oxygen enabled humans to breathe without transporting it from Earth, thus sustaining life on the Moon. Moreover, the colony was able to create the fuel required by the Starship lunar lander (the second stage of the SpaceX rocket) to return to Earth. The hydrogen is easily converted to liquid methane, and the oxygen cooled to produce liquid oxygen as the fuel's oxidizer. The excess methane fuel and liquid oxygen oxidizer are transported by the Starship to the Deep Space Gateway in lunar orbit. Its name and logo are associated with the American frontier symbol of the St. Louis Gateway Arch.

Physically and mathematically, Gateway was not really placed in lunar orbit, but rather an orbit around the Earth-Moon Lagrange point, L2. As a result, Gateway appears to be a highly eccentric, near rectilinear polar orbit around the Moon. It is actually in a halo orbit around the L2 point with a period that varies between six and eight days. Its closest approach to the Moon, periapsis, is 1,900 miles above the lunar North Pole; its apoapsis is 43,000 miles over the South Pole. This orbit enables minimal blackout in its communication with Earth. It is easily accessed with launch vehicles from both the Earth and the Moon. It can also be used as a relay station for operations from the Moon's back side. But it is inherently unstable and requires fuel to maintain orbit. It is powered by four six kilowatt chemical thrusters and by two 12.5-kilowatt Advanced Electric Propulsion System Hall-effect thrusters which maintain its lunar orbit station-keeping with less than 30 feet-per-second of delta-v per year. Spacecraft control is both robotic and autonomous, using executive control software.

The Canadian Space Agency, the European Space Agency, and the Japanese Aerospace Exploration Agency joined NASA in the construction of Gateway in 2028-2029.

Gateway not only provides an outpost to sustain the lunar colony, but also houses the fuel created on the Moon to replenish fuel for any future deep space missions. Excursions from Gateway to low lunar polar orbit is performed with only 12 hours transit time.

The habitats also solved the problem of surviving in the extremely cold lunar temperature. In addition to the absence of any atmosphere, the 40-degree Kelvin temperature during lunar night presents challenging problems to survival. The habitat domes provide the housing for the infrastructure for power, waste disposal, communications, and radiation shielding.

The first Artemis flight in 2027 was unmanned, depositing essential supplies in preparation for the crewed expeditions. Each of the next two missions were manned. Each delivered a seven-person crew of colonists in addition to more supplies. They manufactured bricks from the plentiful lunar dust silicate (regolith) together with an additive from Earth using 3D printers to form the thin building bricks. They built the habitats and connecting tunnels between them from the bricks and covered them with more lunar dust to protect the humans from the ionized solar wind and radiation. The colony was able to extract oxygen from the regolith to replenish the oxygen brought from Earth. The regolith also provided the basis for growing food. Human manure and urine added extra nutrients to the lunar soil to enable the growth of radishes, tomatoes, peas, chives, leeks, cress, and rye. Thus, the colony was self-sustaining, not reliant on additional supplies from Earth for sustenance. But new supplies did include some meat and more seeds, as the new plants germinated slowly.

NASA selected Owens as the navigator for the third crewed mission. The commander for his flight was Andrew Dickson, a 36-year-old, senior Space Force pilot. The rest of the crew were civilians with specialties needed by the colony: medicine, agronomy, engineering, construction, and psychology. One was Andy's wife, a personable elementary school teacher. The colony was beginning to start families, and she looked forward to her new role. Andy expected to retain a leadership role in the colony. The civilians onboard would not have significant roles en route. The Starship primarily served as their means of transportation to their new home. After 15 months of training for the mission, they were all qualified to go.

They were awestruck as they gazed up at the 395-foot stack on the launch pad. It was thirty feet higher than the historical Saturn V which transported the nine Apollo crews to the Moon in 1968-72. The stack was assembled in the enormous Vehicle Assembly Building three miles from the launch pad, then transported on a gigantic trawler to pad 39A.

Six of these historical Apollo crews completed landing on the Moon and invaluable explorations. Among other discoveries, the Moon rocks revealed that the majestic, beautiful Moon was born from a planetary collision with Earth during the first two billion years of formation.

Dickson's crew ascended in the launch tower gantry's elevator to the entry port to their new vehicle home. Each was nervous as they settled into their couches in the SpaceX Dragon capsule atop the Starship stage and the Falcon Super Heavy booster on the launch pad. The 33 Raptor engines in the booster were fueled by cryogenic liquid methane fuel and liquid oxygen oxidizer. Their new spacesuits were more comfortable than the bulky suits donned by the Apollo astronauts 50 years earlier. But still they were looking forward to the shirt-sleeve environment enabled after achieving orbit.

They had a short launch window because they had to rendezvous with another Starship (the Tanker) in low Earth orbit that contained their additional fuel for the translunar burn. The launch crew at the Cape was ready. The Starship crew was ready. The ground controller crew at the Johnson Space Center was ready. They would assume control from the launch control crew in the bunker at the Cape when the booster cleared the gantry. The countdown seemed to take forever. Owens couldn't help but think of the launch explosions in the early years of the space program. He and the others were comforted by the dozens of successes of SpaceX launches after the first three failures in 2007 and 2008. The launch escape system designed to carry the Dragon capsule from the stack in the event of a fire or explosion also gave comfort. It differed from Apollo in that the escape rockets are built into the base of the capsule to push it away from the stack, rather than pulling it away. This made it not only recoverable, but reusable.

Although there was some weather instability in the area around the Cape, they were finally receiving the final countdown. The SpaceX stack's first stage Raptor engines ignited with a thrust of seventeen million pounds, easily lifting the 440,000-pound payload. Musk liked to refer to the stack as the "Big Fucking Rocket," although the media cleaned it up to "Big Falcon Rocket." Thousands of faithful fans watching the launch at the

Cape cheered in unison as the impressive stack cleared the tower just ten seconds after liftoff. As the giant behemoth BFS accelerated into the atmosphere, the crew were pushed hard into their couches, fully experiencing the 4-G force of the powerful engines. The launch went smoothly at first. But 62 seconds into the launch, at an altitude of 27,000 feet, lightning hit the Starship second stage. The displays on the Dragon console lit up like a Christmas tree, then went blank.

Andy excitedly announced, "Houston, May Day! May Day! Alarms! We lost our displays."

"Roger, Starship," the Capsule Communicator at JSC responded in a shrill voice. The backup guidance system for the booster automatically took over control of the engines.

The Capcom continued, "We think it was a lightning strike, knocking out your console and perhaps your guidance system. You are now under control of the backup guidance system. Your ascent appears nominal. Stand by."

One flight controller remembered a similar occurrence during the Apollo 12 launch in 1969.

"Starship, turn on the auxiliary power switch," barked the Capcom.

"What?" responded Dickson, not understanding how the auxiliary power was related. He executed the order.

"It worked, Houston. Our displays are back on," breathed a relieved Dickson, "but we've lost our guidance platform."

"We can restore it," chimed in Owens. "But it will take time. We'll have to wait until we reach Earth orbit."

The first stage engines shut down three minutes into the boost as planned, and after the separation, the second stage Starship's three Raptor engines ignited to complete the burn to orbit.

The spent first stage descended through the atmospheric drag and used three of its engines to touch down perfectly on the launch pad—a feat it had performed many times before, thus enabling the booster to be reused for another launch.

The Starship second stage completed the orbit insertion at 17,800 miles per hour when the Raptor engines shut down as scheduled. Kevin loved floating above his couch, basking in the feeling of weightlessness. He wanted to sing, but duty called.

Houston interjected, "Starship, your orbit insertion state vector looks good. Owens, let us know if you need our help in aligning the guidance platform."

"It should not be a problem, Houston. You guys trained me well. I know what to do."

The civilian crew members unstrapped their seat harnesses and floated up to the cupola observation window to enjoy the topographical sights of Earth from orbit. They all marveled at the sight of the beautiful blue Earth and the white clouds decorating its surface. They noted the vast Atlantic Ocean slowly disappearing as their vision now focused on the European and African Continents. Subsequently they reached the view of the gigantic Pacific Ocean. All the while Owens was aligning the navigation platform.

Owens began using the onboard sextant to lock on to pre-programmed star and horizon sightings to orient the Starship's inertial platform. This had to be done before the first orbit was completed to guide Starship for the translunar insertion burn.

"Great job, Kevin. Your platform looks perfect," congratulated the Capcom. "You saved the mission." Kevin beamed as Commander Dickson echoed the plaudits.

But there was no time to bask in glory. They were approaching the point in the orbit to rendezvous with the

Tanker Starship. It was now visible, orbiting just 10,000 feet in front of them and eight hundred feet above them. Dickson easily closed the gap using the attitude control thrusters, then skillfully executed the docking maneuver. Kevin was impressed. Dickson mated his spacecraft with the Tanker's fuel hose, actuated the valve and began refueling. He remarked that this fueling was a lot easier than refueling the F-15. Upon completion, he closed the valve and slowly undocked. The flight computer identified the point in their orbit to begin the translunar insertion burn. The computer then reignited three of Starship's six Raptor engines, producing 3.1 million pounds of thrust. Almost four hundred seconds later the engines shut down, terminating when Starship reached 24,700 miles per hour. The crew could now relax for their 66-hour coast to the Moon. Once again Kevin was ecstatic, enjoying the effects of Zero-G on his body. He joined the others in the Dragon's cupola to marvel at the beautiful blue Earth as they began to coast away. But the translunar flight was also breathtaking. After four hours, Dickson performed a midcourse correction to ensure the spacecraft was on the free-return trajectory which would loop around the Moon, automatically returning them to Earth in case they could not restart the engines behind the Moon, much like what was required for Apollo 13.

Each crew member worked rigorously in turn on the onboard exercise Peloton equipment. This was necessary to not only maintain their muscle tone, but also their bone mass.

Their Starship steadily lost speed as they receded from Earth's gravity until they reached the Moon's sphere of influence where the lunar gravity began to dominate. This was near the point that the Apollo 13 explosion occurred, aborting the mission and heroically finding a way to return the crew to Earth with no loss of life. The memory of this ominous near catastrophe of

1970 was not lost on Dickson's crew. The transition to the domination of lunar gravity occurred 40,000 miles from the Moon when their speed was down to 2,550 miles per hour. Then their speed steadily increased to 4,100 miles per hour due to lunar gravity as they approached the leading side of the Moon. By passing in front of the leading edge of the Moon, the Moon's momentum transfer helped to reduce their speed. This was the inverse of a gravity assist. It then required only a short burn to further reduce their speed once again to 2,000 miles per hour, enabling capture in lunar orbit. It was during this burn that Starship's engines also performed a plane change to cause the orbit to pass over the intended landing site near the South Pole where the colony resided.

"We have achieved lunar orbit," said Commander Dickson. "Get into your spacesuits, buckle up, and prepare for landing." Then Starship began its final burn to deorbit and land upright and tail first. The three engines which were able to throttle down performed perfectly, resulting in a slow landing on a flat pad less than 3,000 feet from the colony but below the camp's elevation. It had to be located far enough from the colony to inhibit high-speed debris from the engines from spraying surface particles on colony equipment and the colonists.

NASA had employed the early lunar orbiters to examine the many permanently shaded regions to identify those landing sites with easy access to the frozen ice. Each region was a potential "watering hole." The chosen site for the Artemis colony is located near the Faustini Crater quite near the South Pole. The South Pole region is the optimal choice for three important reasons: 1) It provides almost continuous sunlight due to the geometry of the Sun's angle, albeit at an oblique angle to the rays, to enable power generation for the habitat via solar panels. 2) Its vantage point allows continuous line-of-sight communication with Earth. 3) The region has the greatest tonnage of water ice in the many crevices and shadows of its craters. NASA had also considered the Cabeus Crater just sixty-two miles from the South Pole which has the largest tonnage of subsurface hydrogen deposits. But the abundant frozen water source at Faustini took precedent.

The crew slowly exited the ramp to the lunar surface, along with their cargo from the cavernous Starship cargo bay. This bay consisted of the rover, the 3D printers, the solar cell power panels, the foodstuffs, including meat and new seeds, and materials to enhance colonization. Kevin enjoyed bounding across the lunar's one-sixth gravity surface as he headed to the colony's habitats with the rest of Andy Dickson's excited crew. They all entered Habitat 1's airlock, except Kevin. Andy directed him to Habitat 2.

Kevin was impressed by the spaciousness and comfort of the dome. The NASA simulation did not do it justice. After he exited the habitat airlock, he was met by Emily Gardner, a nice-looking woman in her late thirties with short dark hair. "Hi Kevin, I'm Emily and I'm assigned by NASA to be your mentor for the next 30 days. Go ahead and take off your spacesuit and get comfortable. I'm to help you learn all you can about our colony. But they didn't tell me much. I am curious as to why you will be here such an abbreviated time."

"Happy to meet you, Emily. I look forward to learning about your successful colony. But NASA and Space Command have kept me in the dark too. I really don't know why I'll only be here for a month. And I don't know what they are planning for my next assignment. They may be putting me to a test. Or maybe they are planning to start a new colony at the North Pole of the Moon. But I don't know why this would be a secret. As an officer, I am expected to just do as I am ordered. They said they would answer my questions eventually. For now, it is a mystery."

"Kevin, I read the synopsis of your background and I should tell you I am very impressed. I'm sure the powers in charge have big plans for you."

"Emily, you have me at a disadvantage. I'm afraid I don't know anything about you."

"Sure. I am also on loan to NASA from Space Command. I am a lieutenant colonel and was commander of the second Artemis mission to the Moon last year. Since then, I have paired up with Tom Harkins as a colonization couple. We expect to get married next month. For now, I am the deputy commander of the colony, but also the colony's oldest member."

"Well then, I should be saluting you, Colonel."

"No. As colonists we don't worry about rank. Emily will work fine.

"You will notice that we take weekly trips with our rovers to collect the water ice in the shadowed craters nearby for life support. Due to the near horizontal angle, these regions are never in sunlight. With no atmosphere to create moderation, temperatures in these craters are extremely cold, reaching minus 414 degrees Fahrenheit. So, we wear specially insulated suits on these excursions. But, since we are also near Shackleton Crater, there are extremes that are bathed in constant sunlight at an extreme angle for over two hundred Earth days. This enables us to harvest and store the sunlight energy for the dark periods. And the average temperature here is eight degrees Fahrenheit. Our spin axis is 88.5 degrees from the plane of the ecliptic.

"We generate ninety kilowatts of power from our solar panels to sustain everything from life support, to running our scientific equipment, to water recycling. Excess power is stored in batteries for the short durations of insufficient sunlight. We considered using miniaturized nuclear fission for our energy source, but its weight and risk of spreading nuclear material on a pristine landscape nixed this idea. We also considered radioisotope thermal generators. But this would require many devices due to their inefficiency at converting heat into electricity."

"How do you supply food for all your inhabitants?" asked Kevin.

"While some meat products are supplied from Earth, we primarily rely on a plant-based diet. At first, we thought we could grow all our plants in lunar soil. We knew that tomato and wheat could germinate in the regolith. But the absence of nitrogen presented a significant challenge. Also, elevated levels of metals, such as aluminum and chromium can be toxic to plants. Our solution was to use hydroponics, growing our plants in water, rather than soil. This requires copious amounts of water. So, we recycle it from shower runoff, sweat, and urine. Doesn't sound too delectable, does it? But trust me, it works well. The abundance of water ice here makes this an excellent supplement."

"Sounds ingenious, if not mouth-watering," quipped Owens.

"As you can guess, we have to be medically self-sufficient. We have transported all necessary medical equipment needed by the trained medical personnel who were part of the first crew.

"But today, let's begin with helping your crew construct Habitat 3. We'll first make bricks from the lunar regolith using your 3D printers. Three members of the first Artemis mission here are from the European Space Agency. They taught us how to make these bricks after many experiments on Earth. They are great building blocks, not only for habitats, but for roads, and other infrastructure. The resulting igloos are very stable and after we pour more lunar dust on top, they provide excellent protection from radiation while helping to protect us from the cold outside temperatures. Our solar radiation panels generate the necessary heat for our habitats. Although you need your spacesuit outside, we can be in a shirtsleeve environment here in the domes and the tunnels.

"We extract oxygen from the regolith as well since it contains 40% oxygen. We also make fuel and oxidizer from the soil to power our Spaceship's Raptor engines. We use photosynthesis to electrolyze water extracted from the Moon's regolith into hydrogen and oxygen. Then we combine our exhaled carbon dioxide with the hydrogen to produce hydrocarbons including methane, the needed rocket fuel."

"Very industrious," said Kevin.

"The critical resource for the colony is, of course, water. We need ½ gallon per day for each colonist to drink and one-fifth gallon for the food we grow. We have enough water ice at the South Pole frozen in the permanently shadowed deep craters. We recently discovered water near here outside the craters on the lighted surface in the Moretus region. We harvest and melt it to not only supply us with our daily needs, but also to transform it into oxygen and fuel. We derive water and oxygen from this valuable lunar soil.

"Our Moon is bombarded by the solar wind which, among other elements, delivers hydrogen. It interacts naturally with our lunar soil to produce hydroxyl (HO). This in turn produces water when the Moon's surface is heated by micrometeor strikes. Some of this water evaporates into space, but some is trapped in lunar rocks."

"So, how long can the colony survive?" asked Kevin.

"A long time, maybe years. We know we have millions of tons of ice at the North Pole if we deplete the ice here. The issue is that the ice is not replenished at the present time. So, this resource does have a limit."

"Do you get all your power from solar panels?"

"No. We do get some power from the Sun but being near the South Pole limits this. We use fuel cells and nuclear batteries to augment the power we need to heat and light our habitats."

"What about your government? Do you and the commander make all the decisions for the colonists to follow?"

"No, so far strict control is unnecessary. We are all professionals motivated to make the colony a success. Loose government is the key. If we expand significantly, we may need some sort of democracy. One problem is the same as on Earth. With our different cultures and diverse ethnic groups, it is often a challenge to achieve agreements on many issues, both important and mundane."

"I understand you built these habitats and adjoining tunnels in layers from bricks made from regolith and formed from the 3D printers. Are they robust?"

"Absolutely. ESA's procedures have proven to be excellent."

"What countries have joined the colony so far?"

"The United Kingdom has two here. Germany and Japan each have one. India wanted to participate but hasn't so far. The rest are from America. Russia and China are planning to develop their own colonies. Seems foolish to me. But each has their own objectives to mine lunar raw material and harvest valuable lunar resources. We speculate they may have other unannounced objectives. I hope we do not get into parochial conflicts like we do on Earth."

"You spoke of problems due to the diversity and diverse backgrounds of the several countries represented in the colony. Can you elaborate?"

"Well, yes. Our different political backgrounds and religions do sometimes cause tension in setting priorities and making decisions. But so far, we have been able to set these differences aside for the good of the colony. The diversity does get in the way of solidifying friendships, however.

"Our communication with Mission Control is challenging. Not because of the three-second round trip time lag, but because the geometry from the South Pole results in intermittent contact.

"Any other questions before we assist your crewmates with constructing their habitat? We have already formed many of the bricks required to begin."

"None right now. But I would like to clean up. This beard is bothering me. Do I have access to a razor?"

"Yes, I think you will find what you need in the next compartment. When you finish, let's suit up and get at habitat construction."

Emily was pleased to see Kevin's handsome face without the four-day growth.

The Artemis 4 crew manufactured habitat bricks from the lunar soil with the 3D printers to create their new home. After only two days, the dome was complete, and they began the construction of the two tunnel passageways from Habitat 3 to the other two.

"Emily, how do you keep the temperature regulated and habitable inside the igloos?" asked Kevin.

"The method is based on the lunar regolith capacity to absorb and store solar energy from sunlight, then release the stored energy during the night. So, we use the regolith as a heat sink in sunlight and a cold sink at night."

"Is much equipment necessary to sustain the system?"

"Not really. The regolith's properties are similar to the soil on Earth, and it has excellent thermal mass necessary for energy absorption, storage, and release for an energy exchange system.

"Kevin, are you aware of NASA's plan to build a lunar space elevator?" asked Emily.

"I know how they work. Tether a cable anchored somewhere on the equator connected to a satellite circularly orbiting the equator with the same period as the rotational spin rate."

"Yes, then we could send product up the cable to Gateway without needing spacecraft propulsion. Just a motorized transporter."

"Cool. When?"

"They are just plans for now. Nothing committed."

Afterward Kevin and Andy's crew received a warm welcome from their colony compatriots, trading stories of their adventures. Then the colonists related the beginning of their lives on the Moon. They came from diverse backgrounds and ethnicities. Kevin knew one of the colonists, Natalie Stewart. He enjoyed her company when they briefly dated during their training at JSC. She was an agronomist and a major force in growing vegetables within the colony habitats. He was hoping to have some additional female companionship during his monthlong excursion at the colony, but all of the colonists had already paired up as was expected to enable the sustainability of the colony. Attractive Natalie indeed had a mate. Two of the females had already delivered babies, while others were pregnant and expecting the arrival of their new lunar citizens.

Emily helped Kevin learn how to use the equipment to manufacture rocket fuel for Starship's return to Earth. The additional fuel would be transported to Gateway.

Kevin extracted subsurface water and ice, then learned to use electrolysis to separate the cold liquid hydrogen and liquid oxygen and to convert the hydrogen into methane. He loaded these into the fuel and oxidizer tanks kept in the shade of the cold lunar surface. Emily also taught him how to capture the oxygen gas for human sustenance.

But three days later inside Hab 2, Emily's fiancé, Tom Harkins, pulled her away. "What is going on here?"

"What do you mean, Tom?"

"I see how you look at that young, handsome new stud. And I notice you've increased your makeup. I don't want you seeing him."

"Tom, what is wrong with you? NASA asked me to mentor him and teach him about our colony."

"Okay, but now he has learned all about us. So, stay away from him."

"Tom, I can't believe you are jealous. He must be more than ten years younger than I am."

"That makes it worse. Just stay away from him and tell him to stay out of our habitat."

"He'll be gone in a couple of weeks. Stop acting so childish."

"That's not soon enough. Just stay clear of him. You are acting like a cougar."

"**Cougar?** Now you're pissing me off. Stop it! I'm disappointed in you. Trust me. You have nothing to be jealous of."

Emily had to admit to herself that she was a bit attracted to Kevin. But she was no cougar.

***

Kevin would have one passenger on his return to Earth: Jack Chen, who had broken his femur in a habitat construction accident. The resident surgeon had tried to insert a rod in his thigh, but it was not sufficient to give the leg proper strength and it began to bend. He knew it might not work in Earth's gravity but thought it would suffice in the sparse lunar gravity. The surgeon at JSC planned to replace the rod with titanium.

So, after 30 days on the lunar surface, Kevin and Jack said their goodbyes to Emily, Andy, and the other colonists. Kevin had earned their respect. Except for Harkins, they were sorry to see him go. But Owens was happy to leave the cold lunar surface, even though the habitat's environs were better than he had anticipated. But he was anxious to learn about the mysterious assignment NASA was planning for him.

He would now be the commander of the Starship on their return to Earth. They climbed into the Dragon cockpit at the apex of the newly refueled Starship and easily blasted off from the weak lunar gravity.

Once again Kevin is in his element of weightlessness as soon as the Raptor engines cut off. Although this was a direct ascent, it still would take 65 hours before reaching Earth's atmosphere. At the point of atmospheric reentry, Starship's speed is almost 25,000 miles per hour. Thus, the drag causes an external temperature of almost 2,500 degrees Fahrenheit on Starship's skin. It has a Thermal Protection System for reentry similar to the Space Shuttle but is much simpler. Its heat shield consists of thousands of identical hexagonal tiles fastened to its skin. Like the Shuttle Orbiter, the tiles dissipate the heat, keeping the Dragon's cabin temperature survivable. As the Spaceship begins atmospheric entry, Kevin and Chen experience the increasing force of drag pushing them further into their couches, reaching a maximum of 4 Gs. After Starship slows down to a manageable speed, the guidance system maneuvers the forward and aft fins to steer the craft to an upright landing on the original Florida launchpad. Both Owens and Chen were happy to be back on terra firma.

2. The Mission

When Owens got back to JSC, he was greeted by a prominent triumvirate: Roger Tallin, the NASA administrator, Lieutenant General Mark Finley, Space Force's director for human flight, and Anthony Taylor, the NASA director of human resources.

What does this top brass want with me? he thought. They received his undivided attention.

The general was first to speak. "Congratulations on a suburb mission, Owens. You have now earned the rank of Lieutenant Colonel."

"Thank you, sir," responded Kevin quizzically.

“Great job, Kevin,” echoed Tallin. “We are now convinced we have groomed the right man for a very special mission, which we will now reveal. Please listen carefully to what we are about to say.”

“Thank you. I’ve been waiting to hear about my next assignment, but thus far you’ve kept me in the dark. Why?”

“This mission is dangerous and the primary reason for it must not be revealed to the public. Some consider it an impossible mission. The official objective which we will release to the media is space exploration. As John Kennedy said when he announced the Apollo mission, ‘We do this not because it is easy, but because it is hard.”

“I am anxious to hear all about it!”

“That’s why we are all here,” offered Taylor.

“Kevin, I’m sure you are aware of the challenges facing our planet today,” began Tallin. “But you may not be up on the severity of those problems. The first is the threat of climate change. Some scientists predict a temperature rise of 2.5 to 10 degrees Fahrenheit over the next century due to the dramatic increase in carbon dioxide in our atmosphere and the resulting greenhouse effect. This will cause droughts, tree deforestation, flooding, rising sea level, melting of icecaps, and more frequent and stronger hurricanes. More evidence of climate change includes the migration of polar bears inland in search of food, record levels of carbon dioxide, thawing of permafrost in the Canadian Artic, record elevated temperatures in Anchorage, Alaska, record wildfires in the Artic, and hundreds of billions of the Antarctic’s and Greenland’s ice sheets melting and flowing into the Atlantic Ocean. This will cause more frequent and extreme storms. The rising temperatures will cause more water to precipitate and evaporate. This, in turn, will lead to extreme flooding, more intense cyclones, hurricanes, and typhoons. Climatologists forecast the number of large-scale weather disasters globally to be more than five hundred annually. But also, droughts will increase around the world, drying up drinking water sources and water for crop irrigation. This will have a massive impact on human health and hunger. The United Nations has declared that climate change is the single biggest threat facing humanity. It will cause air pollution, disease, starvation, and displacement. The oceans will absorb more carbon dioxide, making them more acidic and dangerous to marine life, The world is currently losing species at a rate of 1,000 times greater than at any other time in recorded history.

"It is true that a minority of scientists disagree with the methodology of using computer models to forecast the future because they are based on current data and cannot reproduce the past. These models assume the Sun's effect on climate is constant. Modeling climate change two decades ago produced totally erroneous results. They point to Earth's 650,000-year history of seven cycles of glacial advance and retreat caused by small shifts in Earth's orbit causing exposure to various levels of the Sun's energy, Therefore some scientists disagree that humans are causing global warming or even that it is really happening.

"Nevertheless, the United States and many other countries have begun to take steps to mitigate it. But these steps will have no effect, so long as countries like China and India refuse to seriously invest in aggressive mitigation plans. China continues to build more coal-fired plants every year. They began building at least forty-one new coal-fired plants back in 2022. Some scientists believe that volcanos and other natural phenomena override any good by us to counter climate change. Thus far, the Paris Climate Accords have had no measurable effect.

"In any event, unless something is done globally, the world and its population is at risk. And of those scientists that believe natural phenomenon, including the Sun, are the cause, then nothing we do will change our destiny on Earth."

Kevin's eyebrows rose. Then his face furrowed into a frown.

Administrator Tallin continued. "Next, humanity is at risk of being wiped out by collision with a large asteroid or even a meteor. Sixty-six million years ago, most of life, including the dinosaurs, was destroyed by an asteroid six-to-nine miles wide which slammed into the Yucatan peninsula at 30,000 miles per hour with a force seven billion times as powerful as the Hiroshima atomic bomb. It created an impact crater 110 miles wide. It hit a subsurface platform of carbonic rock two miles thick, creating the Chicxulub crater. This rock vaporized, releasing an estimated 1,400 gigatons of carbon dioxide into the atmosphere. By comparison, humans inject roughly thirty-two gigatons of CO_2 into the atmosphere annually. The instantaneous regional devastation created a plume that enveloped the entire Earth. This blocked the sunlight, turning Earth into a dark winter. Global temperatures plummeted an estimated 45 degrees Fahrenheit. Subsequently, oxygen in the oceans was depleted, devastating marine life. Then global warming occurred over the course of a few hundred to a few thousand years. This may have parallels to the global warming and ocean anoxia being driven by climate change today. Most plants were not able to survive the effects of the collision, and all animals except the tiniest microbes died. It is not difficult to imagine that if we were struck by such Solar System asteroids or comets even larger today, the population would be destroyed.

"Just eight million years after the annihilation of the dinosaurs, an asteroid six miles wide impacted Greenland's ice sheet, creating a 20-mile-wide impact crater. Scientists estimated the force to be several million times larger than an atomic bomb. This also caused devastation and climate disruption, resulting in the extinction of 75% of plant and animal species on Earth. During Earth's prehistoric formation, we know that large asteroids and comets as large as cities have had catastrophic collisions with Earth. Even today, there is an eight-million-ton asteroid, Bennu, which has a slight chance of striking Earth in the next century. Larger ones would produce Armageddon. We still don't know enough about such a threat. Comets are the greatest threat to Earth because they reach our vicinity of the Solar System with velocities exceeding 100,000 miles per hour. They emanate from beyond the Kuiper belt in the Oort cloud. In 1995 we unexpectedly discovered the Hale-Bopp comet which came within ninety million miles from the Sun. It was visible to the naked eye in 1996 and came closest to the Earth in 1997. It was thirty-six miles in diameter. Had it hit Earth, it would have wiped out most species, including all of humanity. Due to its high speed, it is unlikely we could do anything in time to avoid an apocalyptic catastrophe. The Center for Near Earth Objects at the Jet Propulsion Laboratory in Pasadena, California is currently tracking 28,000 large asteroids and comets that are large enough to concern us and lie within 120 million miles from Earth. But for now, although we can identify and track such threats, we are helpless to deflect or destroy them. In 2022 an asteroid four times as big as the Empire State Building flew within two½ million miles from Earth. A few years ago, an asteroid ½ mile wide traveling at 23,000 miles per hour whizzed by Earth just two million miles away. Two weeks later, a bigger one traveling at 25,000 miles per hour sped past Earth. In 2022 we detected a Trojan asteroid at the Earth Lagrange point L4 in our own orbit around the Sun, less than one million miles away. We estimated it to be ¾ mile wide. Although it is currently stable and trapped in place by Earth's and Sun's gravity, at some point it will drift from L4 and become another threat to humanity. We are currently tracking the asteroid Apophis because in 2029 it will be only 20,000 miles from Earth.

"Recently we learned of significant asteroids that exist in the vicinity of the inner planets. In particular, some, the size of mountains, have been discovered between Venus and Earth. Although we don't know of any threatening Earth today, we don't know that they won't become threats in the future."

Kevin was aware of much of the asteroid threat but thought it unlikely.

"But Earth can also be threatened by a collision with comets, both emanating from within and outside the Solar System. Comets within the Solar System orbit the Sun and usually emanate from either the Kuiper belt or the Oort Cloud. One small comet came from interstellar space, completely outside the Solar System and likely emanated from another star. It impacted the northeast coast of Papua, New Guinea on January 8, 2014. Fortunately, it was traveling in the same general direction as Earth's orbit around the Sun so its encounter with Earth was at 100,000 miles per hour. Had it hit us head-on instead, the relative speed would have been 167,000 miles per hour and caused more damage.

"Countless ancient objects impacted the Moon since its formation 4.5 billion years ago as evidenced by its many craters. The likelihood of Earth being impacted is much higher due to Earth's size and mass.

"The movie *Moonfall* depicted the Moon changing its orbit after a massive collision, then threatening Earth. But this is science fiction because no asteroid is large enough to significantly change the Moon's orbit. However, massive meteors impacted the Moon in its formative days more than a billion years ago. These were denser than the Moon itself and imbedded themselves below the surface. A good example was the Apollo 11 landing. Because NASA modeled the Moon as a homogeneous oblate spheroid and didn't account for these mass concentrations or "mascons," Apollo 11 landed 3.5 miles downrange from the planned smooth landing site in the Sea of Tranquility, and instead had to avoid craters and boulders to find the flat landing spot. As a result, Armstrong landed with just 17 seconds of fuel remaining."

Kevin was aware of this and nodded his head.

"We have experimented with the Double Asteroid Redirection Test in 2022 to try to deflect a threatening asteroid by crashing a spacecraft into it, and it was successful. It redirected the small asteroid, Dimorphous, into a different orbit, and its orbital change was more than we had anticipated. But Dimorphous was only 560 feet wide, while the largest asteroids are more than a thousand times that size. The DART collision with such a larger asteroid would be insignificant. We also concluded that smashing such a threat with a nuclear weapon would not work. The pieces would continue their devastating trajectory."

Kevin's eyes widened. "I was aware of large asteroids in orbit around the Sun between Mars and Jupiter but had no idea that Earth could be in such danger."

Taylor took the next lead. "We not only face risk from our Solar System, but from the rest of the galaxy and even the rest of the Universe. Astronomers have observed gamma-ray bursts as large as one-half tera-electron volts of energy when giant stars explode before collapsing into black holes. This event is one hundred times brighter than supernovas. Another source of gamma-ray emissions occurs when neutron stars collide. They release as much energy as the Sun emits during its entire lifetime. So far, these energy bursts occurred billions of light years away and are of no danger to Earth. But if it occurred only thousands of light years away, it would strip the Earth of its ozone layer and cause mass extinction. In fact, scientists believe such an event occurred 450 million years ago, called the Ordovician extinction.

"Of course, we have other risks developing here on Earth. Kevin, the 2020 Covid-19 pandemic taught us the danger looming in China today. We still don't know whether the Chinese Communist Party intentionally released the virus into the world killing six million people. But it certainly demonstrated the horrible potential of biological weapons to destroy humanity. Thanks to Operation Warp Speed, we were able to develop vaccines in time to curtail the number of deaths in our country and the world. Still, over one million Americans died. But modified strains of the Corona virus could be lethal to everyone, especially if it takes too much time to find effective vaccines. Some believe the CCP might have such ambitions."

General Finley described the next threat. "Colonel Owens, our nuclear triad has deterred our major enemies from a nuclear attack due to assured mutually destructive capability. But the threat of a nuclear holocaust always exists to wipe out humanity. Many scenarios exist. For example, if Iran completes achieving their nuclear weapon objective, their first attack would be Israel. If we retaliate as we promised to do, China and Russia could enter the conflict. We would be forced to counter the CCP and Russian aggression. This could easily lead to a global nuclear war, threatening the entire world. It is not hard to envision a terrorist attack in today's unstable world which could trigger a similar scenario.

"As you know, we all flirted with a nuclear war when Russia, under Vladimir Putin, invaded Ukraine in 2022. When it looked like the invasion was stalled, Putin escalated his attack, committing war crimes against Ukraine citizens, targeting their largest nuclear power plant, and verbally threatening to use his nuclear weapons. He launched his latest ICBM in 2022 to demonstrate the Russian upgraded delivery system. Had we shot it down, that would have ended his blustery threats forever. But we were worried we might miss and that would have strengthened his threats. We were also worried such an attack would cause Putin to escalate. So, we missed a terrific opportunity to demonstrate our ability to thwart a nuclear attack. Only the strong gorilla fighting by Ukraine population using NATO weapons resulted in enough resistance that the invasion was unsuccessful. Had NATO succumbed to the temptation to deliver the air protection requested by Ukraine, a nuclear war might have been imminent, according to Putin. Today the threat by the Chinese Communist regime to invade Taiwan could easily trigger a nuclear war. As you know, we are committed to coming to Taiwan's defense if the CCP moves on them."

Kevin was very aware of these threats but relied on the mutually assured destruction scenario to thwart such aggression.

Administrator Tallin took the next lead. "Another possible threat for Armageddon on Earth is remote but not impossible. As you probably know, our Sun will eventually expand large enough to engulf the Earth. But well before this time the Sun will get hotter and cause Earth to heat up. Our Sun has been increasing its luminosity by 10% every billion years. The heat causes our water to evaporate. Think of global warming on steroids. We don't think this will happen for billions of years, but lately the Sun's flares and radiation have become unexpectedly large and dangerous. We don't believe that this is a sign that the Sun is soon becoming the red giant that would result in the extinction of humanity as astrophysicists predicted one billion years from now. But we aren't 100% sure.

"But we have observed a phenomenon called superflares on some remote stars like our Sun via the Kepler telescope. These can be one hundred times as powerful as the Sun's most formidable flares. If this happened here these would not only cripple all our satellites and communication systems but could also damage our atmosphere permanently in ways we cannot estimate. It is possible to catastrophically endanger humanity."

Owens began to shake his head.

"You are aware of the significant increase of solar flares and coronal mass ejections we have been experiencing during the past several years. This has created havoc with our satellites to an extent never experienced before. We can only hope this is just a super active part of the 11-year solar cycle and not a precursor to our Sun's immediate future."

Digesting all this, Kevin finally responded. "Wow. These possibilities paint a grimmer picture than I realized. But what do you want ME to do?"

General Finley answered. "Owens, we need a way to ensure the survival of the human race. We're not saying such devastation is inevitable, but you can offer an insurance policy for the survival of humanity."

Kevin's eyes grew wide.

Tallin continued, "In the past decade, the Kepler telescope discovered more than 2,600 Earth-size planets, some in the habitual zone, in our Milky Way galaxy. For a while, the planet Kepler-22, which orbits a star like our Sun, looked like the best candidate to support life. Others included Kepler-62e and 62f which have global oceans. Of course, humanity needs water to exist.

‘The James Webb Space Telescope has been fixed on the subset of planets in the habitable zone where surface water likely exists. As you are aware, this telescope is providing new information on planets in the Universe. It is located at the Sun-Earth LaGrange point L2 one million miles from here. Its beryllium telescope mirrors are almost ten times the size of Hubble’s. More importantly it can see through the cosmos dust, detecting infrared radiation. This requires its temperature to be kept extremely cold – almost absolute zero Kelvin. It uses a five-layer Kapton-aluminum Sunshield to block out any trace of sunlight. We are beginning to learn more about these planets from the JWST.”

Kevin had been following some of the findings from the JWST.

“The JWST found a world accessible to us that appears to have environmental conditions similar to Earth to support life. Kevin, we want you to lead a mission to colonize this planet. We believe success is possible.”

“Where is this planet?” asked Kevin.

“Have you ever heard of Alpha Centauri?” asked Tallin.

“Of course,” answered Kevin. “It is one of the brightest stars visible to us. But it is more than four light years away.”

“It is 4.37 light years from our Sun to be precise,” responded Tallin. “It is also a triple star system comprised of Alpha Centauri A and B, plus Proxima Centauri. Alpha Centauri A and B are stars like our own Sun but form a binary pair which rotate around their common barycenter in an orbital period of 79 years at a distance from each other varying from one billion miles to 3.3 billion miles. So far, we have not detected any planets around these two stars. But Alpha Centauri C, otherwise known as Proxima Centauri, is a small, faint red dwarf main sequence star located 1.2 trillion miles from the bright pair and 4.24 light years from our Sun. It is not readily visible from Earth without a powerful space-based telescope. In 2016 we detected at least three planets that orbit Proxima Centauri. One of these exoplanets, Proxima b, is in the habitable zone. It orbits Proxima Centauri from 4.6 million miles every 11.6 days. Because it is so close, it is tidally locked to the star, just as our Moon is tidally locked to Earth. Proxima b is 30% larger than Earth but has a mass only slightly larger than that of Earth. We don't yet know much about its atmosphere, but so far, the James Web Space Telescope reveals the atmosphere is composed of nitrogen, methane, and oxygen. We want you to lead a team to colonize this planet. However, being more than four light years away means that even traveling at an average speed of 50% the speed of light, it will take more than eight years to get there. Therefore, it is likely that this will be a one-way mission."

"How did you find this planet?" asked Owens.

"As you may know," responded Tallin, "we discovered over 5,000 exoplanets, primarily with the Kepler telescope. In fact, Kepler data infers there are two billion planets in the Milky Way. We think three hundred million of these are potentially habitable. We narrowed the list of these 5,000 down to several rocky terrestrial planets that were Earth-size, resided in the habitable zone, and likely to have surface water. They have a bulk composition dominated by rock or iron, and perhaps carbon. Astrobiologists hope to study starlight that has interacted with a planet's surface or its atmosphere to determine whether it was transformed by life. We call molecules that are necessary for life to exist a biosignature. For the first couple of billion years, before

Earth's atmosphere even had any oxygen, it hosted simple, single-celled life. But 2.4 billion years ago, a new family of algae evolved. This algae, via photo-synthesis, began producing free oxygen. Ultimately this produced a strong biosignature. We hope to eventually detect the biosignature of oxygen and/or methane in the atmosphere of habitable planets. We can't be 100% sure of their atmosphere or whether they possess oceans, but we can surmise these properties based on their distance from their star, their size, and other data. For now, only the James Webb Space Telescope is capable of such detections. But the JWST was not designed to search for life. It can only detect changes to atmospheric levels of carbon dioxide, methane, and water vapor as the planet passes in front of its star. The telescope must block the bright light of the host star to reveal the light reflected from the planet. As we learn more by analyzing the light from these exoplanets, we can better determine whether the atmosphere and water resemble Earth and whether the temperature is similar. But even then, it will be possible to obtain false positives. For example, volcanoes and even cows on Earth release methane into the atmosphere. Also, sunlight splits water into hydrogen and oxygen. These could be erroneously interpreted as biosignatures.

"Astronomers used both the transit and doppler methods to discover the existence of possible planets that may have these atmospheres. We have planned three enormous new ground-based telescopes to search for biosignatures: the Giant Magellan Telescope, the Thirty Meter Telescope, and the European Extremely Large Telescope. Unfortunately, it will be years before we can detect such biosignatures."

"I am familiar with both methods of detecting exoplanets," boasted Owens. "The transit method detects the drop in luminosity of the star as the planet crosses our line of sight to the star. The doppler method detects a "wobble" motion of the star itself, caused by the gravity of the nearby planet. That's how astronomers discovered Saturn, Uranus and Neptune."

"But our ability to detect life or even a life-supporting atmosphere on planets outside our Solar System is quite limited at the present time," added Tallin.

"Astronomers discovered Proxima b using the Doppler method. The transit method was not possible because Proxima Centauri is too dim to detect any change in luminosity caused by the planet's transit. But Proxima b's close location and relative mass made the Doppler method feasible. If the exoplanet were farther away, not in the orbital plane, or smaller in mass, it would be much more difficult to detect.

"We also discovered seven excellent candidates in the habitable zone of TRAPPIST-1. Some likely have more water and water vapor than Earth in their atmospheres. They appear to be suitable for some form of life. But they are thirty-nine light years away, which is why we focused on Proxima b."

"Back to the voyage," questioned Owens. "How are we going to travel four light years distance in only eight years? It took 36 years for Voyager 1 just to leave the Solar System. How is it possible to reach nearly the speed of light?"

General Finley responded, "We have developed a breakthrough nuclear propulsion engine at Los Alamos Laboratory which is capable of emitting ions, continuously accelerating a spacecraft throughout flight at significant acceleration levels. It was developed based on a theoretical concept, called a Helical Engine published by David Burns at NASA's Marshall Spaceflight Center. They solved the problem of housing the reactor with a material that will not melt at the elevated temperatures required by the nuclear fusion process. We can use the excess heat as a biproduct to keep the spacecraft cabin warm in the cold interstellar space. Continuously expelling superheated ions produce the relevant thrust to weight ratio to accelerate a spacecraft almost indefinitely. So, in less than two years in the absence of gravity, the spacecraft could reach 90% of the speed of light."

Tallin added, "Our simulations show that the Sun's gravity will, of course, have a major effect initially, but using gravity assists from the planets, the spacecraft will leave the Solar System in less than one year. At this point, the Sun's gravitational effects will be small, enabling the nuclear engine's thrust to reach 90% the speed of light in less than two years. Of course, reverse thrust and deceleration will be necessary to slow the spacecraft down for orbital capture by the exoplanet and for landing. This ion engine is currently undergoing final testing prior to installing it in the SpaceX Starship, coexisting with the chemical propulsion engine you just flew. The forty-two chemical Raptor engines will enable achieving Earth escape velocity. This is a significant upgrade from the 33-engine booster you launched to get to the Moon. Then, instead of coasting, we will use the nuclear engine throughout the flight, including braking as it approaches the Alpha Centauri star system, when the Raptors will reignite for final braking and landing."

Tallin concluded, "We are not exactly sure of the ultimate speeds and therefore the duration of the flight, but this will be known as we observe the nuclear engine performance. Of course, at such high speeds relativistic effects are measurable."

"Why all the way to Proxima Centauri?" asked Owens. "I understand why the Moon cannot be permanently colonized due to the water and radiation limitations, but what about Mars?"

"We did consider both the Moon and Mars for this survivable colony, and we may still colonize Mars for an abbreviated time as we did the Moon. Elon Musk has designed a modified version of the SpaceX Starship to colonize Mars with dozens of passengers per journey, eventually reaching a robust colony in the next decade. But both the Moon's and Mars' environments are too harsh for indefinite survivability. We would have to deal with the perchlorate chemicals on the Mars' surface that are perilous to human life. They are chlorinated hydrocarbons that are globally distributed. We observed with alarm the massive buildup of regolith due to dust storms on each of our robotic landers on the planet. Also, Mars is exposed to a great deal of radiation due to its lack of a magnetosphere and thin atmosphere. The shear mass of Earth's atmosphere is 5.5 quadrillion tons. But Mars' thin atmosphere provides much less shielding with a mass less than 10% of Earth's atmospheric mass. Add in the lack of any magnetosphere results in a dramatic radiation danger. The Moon, of course, has even more of a radiation threat with no atmosphere nor magnetosphere. That is why it requires radiation protected housing and protective spacesuits. We believe that Proxima Centauri b may closely resemble Earth with both a strong atmosphere and, hopefully, a magnetosphere to offer strong radiation protection. Maybe even to the extent that a radiation protection spacesuit may not be required.

"We have also considered colonizing one of the largest moons in our Solar System. We looked at the four large Galilean satellites of Jupiter. You know that our Juno probe launched in 2011 began its polar orbit around Jupiter in 2016. Although the primary mission was to learn about Jupiter's composition, gravitational field, magnetosphere, atmosphere, winds, and mass distribution, we learned something about its moons. Io, the innermost moon, is slightly larger than Earth's Moon. It is the most geologically active body in the Solar System with over four hundred active volcanoes. Its atmosphere is primarily sulfur dioxide and therefore not a candidate for colonization. The second moon, Europa, is slightly smaller than our Moon. It is mostly covered with water with a thin layer of ice on top. It may be entirely covered with this water and ice, thereby eliminating it as a candidate for colonization. Jupiter's third satellite, Ganymede, is the largest moon in the Solar System. It is believed to have a saltwater ocean below its surface, which could make it a possible candidate for colonization. Also, its thin atmosphere contains some oxygen, ozone, and carbon dioxide. But its surface is believed to be over one hundred miles of thickly layered ice. Once again not ideal for a colony. Finally, Callisto, the second largest of the four, is slightly smaller than Mercury and likely also has an underground ocean. But its surface is extremely cold – 218 degrees below zero Fahrenheit; too cold for permanent colonization. But some carbon dioxide was detected in its thin atmosphere. So maybe it is habitable, but not for humans.

"But we also looked at Titan, Saturn's largest moon, which is 50% larger than Earth's Moon and almost the same size as Mercury. We received significant information about Titan from our Cassini probe in 2004. At first, its Earth-like features such as lakes, rivers, and seas and dense atmosphere made Titan an exciting candidate. But we now think that the surface features are filled with liquid methane, rather than water. And its atmosphere is mostly nitrogen ammonia, which can be toxic. We were unable to detect any oxygen.

"That takes us to Proxima b. Based on the latest observation from the James Web Space Telescope, Proxima Centauri b has environmental conditions similar to Earth. It has a surface ocean and an atmosphere that promises to warm the planet similar to Earth's temperatures. In addition, it has a small percentage of oxygen mixed with ammonia and helium that may be breathable. Proxima b has matured beyond its immature biosphere stage of almost all carbon dioxide and ammonia. It is becoming a mature biosphere with significant and increasing quantities of oxygen. We believe it now can sustain life at least in plant form. This resembled the Great Oxygen Event, which began on Earth some two billion years ago."

Kevin began to get excited over the genuine prospect of the mission.

General Finley advanced, "Colonel Owen, you were carefully chosen for this assignment, based on your intellect, education, achievements, leadership skills. patriotism, and youth. But we can only give you a few days to decide, as we must get on with choosing the rest of the crew and training them. Surprisingly, we have received over 3,000 serious applications for this mission."

"I can give you my answer right now," said Owens excitedly. "There is nothing I would rather do than lead a team on this odyssey."

"We hoped you would feel that way," said Tallin.

"We will select the crew together," said Taylor. "You, me, and Sheryl Stanton, our chief psychologist. We can begin tomorrow. We have culled the applicants down to the top twenty in order of their qualifications and compatibility. We looked for key traits such as intelligence, accomplishment, collaborator, social adaptability and maturity, tolerance, motivation, and ability to function in dangerous situations, We have given the first twelve a complete physical examination and determined they are healthy as well as capable to procreate. We also performed extensive psychological testing to minimize the likelihood that the candidates would not be able to withstand the pressures of trauma on such a taxing extensive journey. These include such traits as handling stress, sensitivity to others, ability to deal with separation from loved ones, and ability to tolerate isolation. Unfortunately, we had to eliminate three wonderful candidates for whom we were worried might be susceptible to mental anguish.

"Kevin, together we will first select the additional five primary crew members. Then we will proceed to select a backup crew as we always do for our missions. Any questions or comments?"

"I'm ready and excited to begin. I'm sure I will have many questions as we proceed," responded Kevin.

3. The Crew

Prior to the interviews to select the crew members, each candidate completed a questionnaire. NASA used it primarily to determine crew compatibility. This was crucial to ascertain whether the crew could successfully coexist for the rest of their lives, particularly on their more than eight-year journey. NASA learned from the long durations on the International Space Station that culture incompatibility could promote an "Us versus Them" partitioning that could compromise mission objectives. NASA hoped the questionnaire would expose these issues, enabling the exclusion of candidates whose strong cultural values and ideologies were conflicting with other crewmates. Therefore, there were no right or wrong answers to the questions. The first order of business regarding compatibility was to determine the biases and beliefs of the crew commander, Kevin Owens.

The questionnaire included the following:

1. Religion:
a. What is your religion?
b. Do you believe in God?
c. How strongly do you adhere to your religious principles?
d. Do you believe in life after death?
e. Do you believe that the Universe was created by God?
f. Could you marry someone with different religious tenets?
2. Politics:
a. Do you consider yourself politically a liberal? Progressive? Conservative? Independent?
b. Are you formally affiliated with a political party?
i. If so, which one?
ii. How strongly are you aligned with that party?
c. What are your views on the first ten amendments to the constitution?
d. What are your views on law enforcement?
e. What are your views on abortion?
f. What are your views on taxes?
g. What are your views on free enterprise? Capitalism? Socialism? Progressivism? Communism?
h. What are your views on legal and illegal immigration? Asylum?
3. What is your sexual preference?
a. What is your ideal mate?
b. Do you want to raise children? How many?
4. Could you become a vegan? Would you enjoy it?
5. Do you have any anxieties about being in an enclosed space?

6. Who are your closest relatives and friends?
a. How would you cope with their loss of contact?
7. Can you unquestionably accept an order from the spacecraft commander?
8. Do you have any racial prejudices? Preferences?
9. Will you accept all fellow crewmates as equals?
10. Which human traits annoy you most?

Owen's answers were not unexpected:

1. Religion:
a. Christian.
b. Yes, I do believe in God.
c. I do adhere to my religious tenets.
d. I believe in an afterlife.
e. I believe God created the Universe.
f. While I would prefer my mate be of similar faith, I might marry someone whose religious beliefs differ from mine as long as she respects my tenets.
2. Politics:
a. Conservative.
b. I am a registered Republican, but not strongly affiliated.
c. I strongly support the Bill of Rights.
d. Laws are created to be enforced. No exceptions.
e. I believe abortion is wrong, except to save the mother's life or if the impregnation was by rape. I believe the fetus is a human being and as such has rights that must be protected.
f. Taxes are a necessity to meet several objectives but should only be imposed to satisfy these necessary objectives.
g. Capitalism and free enterprise are the backbone of what made America great.
h. I am strongly in favor of legal immigration but just as strongly opposed to illegal immigration. I am also against bypassing legal immigration on the basis of

declaring asylum because it has been so flagrantly abused.

3. I am 100% heterosexual.
a. My ideal spouse is of equal intellectual capability. We have complete devotion to each other and mutual respect. We have a lifelong commitment.
b. I definitely want children, but the number will be negotiated with my wife.
4. I could accept becoming a vegan but might not enjoy it. I love a good steak.
5. I've proven I can thrive in an enclosed space on my voyage to the Moon last month.
6. I lost my parents in an airplane crash. I have no siblings. I have made friends at the Air Force Academy, in pilot training, and my career as an astronaut. But I can't say we have become lifelong friends. So, I can manage the loss of personal contact better than most.
7. I proved I can accept orders on my lunar voyage as well as in my military and astronaut career to date.
8. I don't. I am free of any racial prejudices whatsoever and have proven that throughout my adult life. No preference.
9. I will accept my crewmates as equals, except where necessary to assert my authority to make decisions as their commander.
10. I am annoyed by laziness and not giving one's best at all times.

NASA loved Owen's answers and they reinforced their conclusion that that they had selected the right stuff.

***

Each candidate was given a thorough physical, mental, emotional and psychological test prior to his/her interview.

Anthony Taylor began each interview with the same introduction. "You have been invited to meet with us due to your outstanding qualifications and interest in this unique mission. As you know this is a trip from which it is likely you will never return to Earth. Although there is risk, we believe the mission will succeed. SpaceX has developed a booster and spacecraft specially designed for this mission. You will travel to an exoplanet more than four light years away which we believe has an environment similar to Earth's, supporting colonization of indefinite duration. But we estimate a flight of approximately eight years duration, maybe more. Although the newly modified spacecraft has ample size for privacy and separate pods for quarters, you will be in proximity to your fellow crewmates for the rest of your lives. Therefore, it is important that you are all compatible. This committee will evaluate not only your qualifycations to execute your role, but also your persona as a crew member.

"This committee is comprised of your spacecraft commander, astronaut Lt. Colonel Kevin Owens, NASA psychologist Sheryl Stanton, and me, Tony Taylor, the NASA director of human resources. In addition, your fellow crewmates will help in the selection.

Any questions before we begin?"

The first applicant is Dr. Jennifer Thorson, a stunningly attractive surgeon and obstetrician from Silver Springs, Maryland. She sports long blond hair tied into a ponytail that highlights her pretty face. Her makeup is minimal. She wears a dress that accents her curvaceous body. Her natural beauty stunned Owens.

Tony asked, "Dr. Thorson, do you feel qualified to perform any surgery which may be necessary on this mission?"

"Please call me Jennifer. Although I'm only a second-year resident, my success in supporting and leading a variety of surgeries at John's Hopkins, one of the best hospitals in the country, qualifies me for both routine and emergency situations, as my colleagues will attest. In addition, although I'm only 25 years old, I am also proficient at obstetrician practices including deliveries. I understand you do intend to populate the colony, and my skills will be essential for mission success."

"Why do you want to undertake this challenging mission?" asked Kevin.

"Ever since I was a child, I've always been fascinated with space travel. My medical training has always been most important, but deep down I knew I would jump at an opportunity to combine these interests."

Kevin began scribbling copious notes. He knew he had to get past his physical attraction to her to concentrate on her as a professional and a crewmate.

"What about your family and friends?" asked Sheryl. "This mission means you will never see them again."

"I have not really had time to develop close friends apart from my colleagues at Johns Hopkins," Jennifer admitted. "I have never married, have no siblings, and my elderly parents support my career decision if this is what I want. Yes, I would miss them and they me. But they understand how important this unique adventure would mean to me."

Sheryl offered, "At this point we have to explore some very personal subjects. I hope you understand this is necessary to ensure these issues will not create problems in your compatibility and relationships with your crewmates. Jennifer, what is your sexual orientation?"

"I am and have always been heterosexual, although I am not very experienced. As I stated earlier, my medical education has limited my time dating. I have enjoyed being with male colleagues and have become intimate, but never reached the point of considering a permanent relationship. Down the line, I would welcome a permanent relationship, like to marry, and raise a family. But not yet."

"There may be a great deal of time with virtually no duties on this mission," said Kevin. "How would you propose to occupy your time?"

"First, I would look forward to developing a strong relationship with my crewmates. This is something that I realized has been missing in my life. What an opportunity to satisfy a longing without competing demands on my time which precluded such relationships in the past.

"Obviously, I have thought this through. I also would take advantage of hobbies for which I did not have time. Reading for pleasure, exercising, learning card games, even watching television would be welcome diversions in my life. I would not be bored."

"What are your political views?" asked Taylor.

'I suppose I do have strong convictions on a variety of subjects. As you might have guessed, I could be characterized as a political conservative. I believe people should apply their special talents and education to contribute and thrive in this world, regardless of their station in life. I think our government should not be expected to support them. I believe government handouts have resulted in dependency, thus inhibiting personal growth and self-esteem. In my opinion, this has promoted generations of perpetual absence of self-reliance. I believe some of our politicians created this government dependency to have power over their constituents, thus promoting their re-election.

"I apologize for this rant, but I think it is important to let you know these views now in case this would be at odds with my fellow crewmates. If any of the other five have opposing viewpoints, I think this could cause a problem in our relationship and mutual respect."

Kevin was now admiring more than her beauty. He tried to be more professional and asked, "With your taxing schedule, have you been able to participate in any hobbies?" asked Kevin.

"Well, as I said earlier, I have always wanted to learn to play cards but never found the time. I even downloaded several books on contract bridge and poker but haven't even read one of them so far. I should find the time on this incredible journey to do so, don't you think?"

"Absolutely," answered Kevin. "Maybe our entire crew could develop this hobby."

"Are you religious?" asked Sheryl.

"I am, but probably not devout. I believe in God and was raised a Christian, but attend church only on rare occasions, such as Christmas and Easter."

Taylor stated, "We appreciate your candor, Jennifer. Now, do you have questions for us?"

"You do expect me to choose a mate on this journey, right?"

Tony responded, "That is our hope. We expect to populate this colony and you will be key to that objective. We will be selecting three women and three men for the crew in hopes you six will pair up to begin to populate this new world. We would like assurances that you do not start your families until you land; we do not think it a good idea to add babies to the Starship passenger roster. To that end we will supply both male and female contraceptives."

"Will I have a say in who my crewmates are?" She smiled demurely at Kevin.

Kevin's heart started beating a little faster. He wondered, *Is she flirting with me? Nice!*

"Definitely. If you are selected, you will be part of the selection committee."

"If there are no other questions at present, we will get back to you shortly," offered Taylor. "If you have further questions, we can answer them at that time."

After Dr. Thorson left the room, Kevin was first to comment. "She is perfect. What a great crewmate she would make. We need her expertise." He did not mention that he was so attracted to her beauty and figure.

"I agree," said Taylor. What do you think, Sheryl?"

"She appears very stable, confident, and compatible. I vote yes, but we need to check her references more thoroughly."

After the intense vetting, they all agreed to invite Dr. Jennifer Thorson to the team. She accepted immediately and was ecstatic. She said she would only need a week to get her things in order and give the hospital notice.

Taylor invited her to the interview for the next candidate: Michael Stone, a 27-year-old engineer from Madison, Wisconsin. He displayed a striking picture of a confident, six-foot, muscular-toned, blond Adonis with a square jaw and chiseled features. He was dressed in a subdued sport shirt and slacks that made his hard body evident.

After Taylor gave his standard introduction, he introduced Owens, Sheryl Stanton, and Dr. Thorson to Stone.

"Michael, let's begin with your qualifications."

"My undergraduate degree at the University of Wisconsin was a double major: civil and mechanical engineering. I went on to earn my master's degree in electrical engineering. From what I know about this mission, I may have the opportunity to use each of these disciplines."

Sheryl asked, "Please tell us about your life."

Michael began, "I had a rough childhood, as my parents did not want a family and surrendered me to the Milwaukee county welfare system before I was a year old. I spent the next 17 years in various foster homes. I discovered my intense interest in figuring out how things worked at an early age. I did have some excellent teachers in high school, who helped me to obtain a full scholarship at Wisconsin. This led to a teaching assistantship in graduate school. I chose EE because I knew I needed this discipline to complete my understanding of how things worked.

I dated throughout high school and college, enjoying the company of coeds at UW. But I didn't get too serious until I left college to begin my engineering career.

I met a girl whom I thought would be my life mate. We discussed raising a family before we married. But soon afterward, I found Christ and became a born-again Christian. Both of us had been agnostic before we were married. My wife did not share my devotion to the Lord. We struggled with this conflict, and it was not resolved when she got pregnant. She did not want the baby in her life yet. She aborted the fetus against my strong objections. We quickly drifted apart and divorced within a year. It took two years for me to recover and start dating again. But I am not in a committed relationship presently."

Kevin asked, "Is religion an important part of your life today?"

"Very much so," Michael responded. "I am currently a deacon in my church."

"If any of your crewmates were not Christians, would that be a problem in your mutual relationships?"

"It would be in choosing my life mate, but not with the rest of the crew. Some of my current friends do not share my faith, but we have no problem setting those differences aside. Christ says to love your neighbor as yourself. I subscribe to that teaching."

Jennifer requested, "Michael, please tell us about your political views."

"As a teen, I was more liberal than I am now. But when I saw progressivism evolving into socialism, I changed my view. I love this country and hate to see the self-reliance that gave us the productivity that made our country great be undermined by government dependency. I believe the government should get out of our way to enable individuals to pursue their dreams the way the founding fathers intended. I am upset with our politicians that put their re-election self-interest ahead of what is good for the country at the local, state, and federal levels. And lately, the lack of fiscal discipline has created debt that cannot be sustained."

Jennifer was impressed, not just with his looks but also his political bent. Kevin nodded his head in agreement.

"The crime wave that is sweeping the country lately is also the fault of the politicians. Their effort to defund the police has caused our law enforcement officers to leave the force in droves, enabling crime in our streets to run rampant. It is not safe to be out there. And our criminal justice system is failing us, putting criminals back on the streets, rather than incarcerating them.

"I am sorry I sound so negative, but I believe my crewmates should know how I feel. I want to avoid personal conflicts in the journey and the colony."

Sheryl followed up. "Michael, why do you wish to take this bold step?"

'Space is so fascinating to me. I took an elective course at Wisconsin in astronomy, and the fascination has never left me. We had a 15.6-inch Clark reflector telescope on Observatory Drive called the Washburn Observatory. We viewed the Moon, the Galilean satellites around Jupiter, Saturn's rings, star clusters, and four galaxies. Exploring space would be the adventure of a lifetime. And to explore a distant planet would be mind-boggling. I could not ask for a better way to spend my life."

Taylor said, "If the Starship spacecraft is as reliable as we hope, there may be a great deal of downtime for you until we land. How would you spend this time?"

"First, I believe there will be much to see on the way out of the Solar System. Then in extrasolar space I would expect to be building a strong relationship with my crewmates. Also, I have always been an avid reader. I plan to store thousands of eBooks on my Kindle. My favorites are John Grisham novels and science fiction. But I also read nonfiction biographies related to space history. I also enjoy daily exercise to maintain and improve my body. I was a varsity wrestler in high school and in college. I hope we will have state-of-the-art exercise equipment onboard. Finally, I would hope to find mutual entertainment activities to do with the crew. I would hope to fall in love and marry again."

"Do you have other hobbies?" asked Jennifer.

"Yes. I discovered I enjoy scuba diving, although living in Wisconsin makes this opportunity rare. Come to think of it, this will be rarer still in interstellar space. But maybe our exoplanet will have an ocean to enable this hobby."

Tony concluded, "Michael, I'm sure you have questions for us."

"Yes, I do. I don't understand what kind of engine can get us to the exoplanet in only eight years. Also, what food, water, and oxygen will we have on board for such a duration? Finally, who are the crew?"

Kevin responded, "You are looking at two of the six crew members, Dr. Thorson and me. The others are yet to be selected. But if you are one of the six, you will have a say in their selection.

"NASA is building into their new Starship an amazing new nuclear engine along with its primary Raptor chemical engines. The nuclear engine will provide most of the long-term acceleration required to obtain these great speeds."

Tony continued, "NASA is modeling our food supply after what they learned on the International Space Station. We will have more than a year of prepackaged food onboard, much like both the Space Station and the lunar colony. But we expect to grow our own vegetables and possibly some fruit hydroponically. In addition to the water that we bring onboard, NASA has perfected a system to extract drinkable water from our urine. They also have devised a unit to separate water into hydrogen and oxygen as was demonstrated in the lunar colony. Once you land, you will extract water from the oceans we believe we detected on the exoplanet. We expect you to grow vegetables from the regolith on the exoplanet, much as they do in our lunar colony."

"What if we encounter an irrecoverable problem preventing us from reaching our destination?"

"We expect to have enough fuel to return to Earth and land."

"Will I be trained in the maintenance and repair of the assets of our spacecraft, including the engines?"

"Absolutely," answered Kevin. "You will have a year of training to become proficient with these amazing systems."

"Anything more?" asked Tony.

"Not for now, but I may have other questions later."

Taylor responded, "Of course.

"We will be in touch with you on our decision soon."

After Stone left, each member of the committee was polled. Both Kevin and Jennifer were satisfied, pending the vetting. Jennifer suggested quizzing the references as to whether his strong religious convictions could produce conflict with other crew members. Sheryl and Taylor were quite impressed.

After the reference checks came back with a glowing report, Taylor called Stone to invite him to the team. He enthusiastically accepted and volunteered to begin astronaut training after giving notice to his employer. He was happy to be invited to the selection committee.

***

The next candidate was Ashley Stevenson, an attractive 22-year-old botanist with long auburn hair from Amarillo, Texas.

"Welcome Ashley," greeted Tony. "I am Tony Taylor, the NASA director of human resources. Next to me is Sheryl Stanton, a NASA psychologist. The other three are crewmates selected for this mission: Lt. Colonel Kevin Owens, the mission commander, Michael Stone, our flight engineer, and Dr. Jennifer Thornton, the flight surgeon.

"As you have learned from our request for applicants, this is a highly unusual, one-way mission. The crew of three men and three women will travel in space for more than eight years to colonize a planet outside our Solar System. We know only from our telescopes what this planet may be like. It is in the habitable zone of a red giant star. We know the planet has sufficient water, even oceans, to sustain life. Although it is larger than Earth, it has nearly the same mass. Therefore, gravity is slightly less than Earth. We believe that near the proposed landing site the temperature will be similar to Earth's. The big unknown is its atmosphere, which we can only surmise from our telescope observations. It should have clouds and water vapor if we are right about its oceans. But we have only detected lesser amounts of oxygen and carbon dioxide. The CO_2 indicates it likely has plant life."

Sheryl continued, "In part this interview is to determine whether you will be compatible with the crewmates for this extraordinarily lengthy journey and the colony itself. Therefore, we may be asking questions of a very personal nature.

"We want you to help colonize this planet, raise a family, and sustain life indefinitely. So, Ashley, are you up for this challenging mission?"

"Yes, of course. This is why I applied."

"What contribution will you make to meet our objectives?" asked Kevin.

"Well first, I am young, healthy, and fertile, as your medical evaluation has proven. Next, my botany degree will be invaluable for growing food in the colony."

"Won't you miss your friends and relatives?" asked Sheryl.

"Probably. But my relatives are mostly dysfunctional, and we haven't been close since I went to college."

"And your friends?"

"I met a lot of people in college and in my field, but I can't say I am very close to any of them. My college roommates didn't last more than one or two semesters."

"Please tell us about your sexual orientation and experience," requested Sheryl.

"I am strictly a straight female and like men very much. I have dated men throughout college and since graduation. So far, these dates have not gone beyond the point in which we got very serious."

"Please tell us about college," asked Jennifer.

"Well, as my resume indicates, I first enrolled in Texas A&M in farming. My sophomore year I transferred to North Central Texas, so I could save money by commuting from my parents' home in Gainesville. I earned my bachelor's degree in botany."

"Do you have any strong views about politics related to local, state, and federal government?" asked Jennifer.

"Not really. I changed my party affiliation from Democrat to Republican, then to Independent. I don't like the strong emotions and fighting that goes on in politics. I think we should all get along. Political discussion and arguments make me uncomfortable. I avoid them."

"Are you religious?" asked Michael.

"Not really. I believe in God. So, in that context I guess I am religious. But I have not chosen one religion over another."

"What attributes would be important to you in choosing crewmates?" asked Kevin.

"Well, they should be pleasant and kind and appreciate me as a person."

"Do you have any question for us?" asked Taylor.

"Yes. What opportunity will I have to select my husband?" She looked directly at Kevin and Michael.

"If you are selected to join the team, you will have plenty of time to get to know your crewmates before pairing up," responded Tony. "The six of you will be in intense training for approximately one year. Then, of course, you will have the entire journey together to make a final decision on a mate. We would hope you could make that major decision sooner, rather than later, but we will not put you under any pressure."

"One more question: Why can't you select a planet closer to home to avoid such a long flight?"

"As you know we do have colonies on the Moon, but they are not sustainable," answered Taylor. "We have considered all other possibilities in the Solar System, but each has challenges for human survival. This planet is the closest that has an environment we believe will sustain human life indefinitely, enabling the permanent colony.

"If you have no other questions, we will contact you within the next day or so with our decision. In the meantime, if you have other questions, you have all our phone numbers."

After Ashley left the meeting, the group began their evaluation comments.

"I was not impressed." exclaimed Jennifer. "She appears so shallow. What were her college grades?"

"She recorded a 2.5 GPA at A&M, then a 2.7 at North Central Texas," responded Tony.

"I think she would not cause any relationship problems," said Kevin. "She seems likeable enough and would probably get along. But I agree with Jen. It's not clear to me that she would make a real contribution to the team."

"Her apparent lack of commitment to virtually anything important is a problem in my opinion," suggested Michael.

"I am frankly surprised she was placed so high in the order of candidates," apologized Tony. "We have so many applicants that appear stronger."

Each of the committee agreed that Tony should call and let her down gently.

Taylor introduced the next candidate, Samantha Gere. She was a pretty, 23-year-old agronomist from Des Moines, Iowa. Her long, brown hair framed her appealing facial features. She flashed eyelashes that did not hide her pretty brown eyes. She wore a tight-fitting blouse and slacks that left no doubt about her impressive figure. Michael was impressed by her looks.

Tony introduced each of the committee members and ran through his introductory message on the mission and the interview ground rules. She looked over the abbreviated crew and liked what she saw. She thought Michael looked quite viral.

"Samantha, please tell us about your background for this mission," Tony requested.

"I grew up on a family farm in rural Iowa. I was always interested in different, modern ways to enhance the growth of food. At Iowa State University, my degree in agronomy included hydroponics and unusual ways to grow food without soil. When I learned the importance of these techniques for producing food in the International Space Station and in the lunar colony, I devoted my studies to this end. As a result, I believe this is exactly what you all need for this exotic mission to succeed."

Owens asked, "Why do you want to use your knowledge for this challenging mission, rather than devising techniques for space application while you remain in a lab on the ground?"

"My interests involve more than my profession. If I understand your mission correctly, it is an opportunity to create a whole new society on this exoplanet. We can begin fresh, without all the baggage brought by current society here on Earth."

Michael began nodding in agreement.

"Please elaborate," requested Sheryl.

Samantha continued, "I am concerned that the Chinese Communist Party will fulfill their prognostication of surpassing the United States in every way and take over the world. And our politicians will not stop them. We already have some government representatives that function as socialists, and our Congress seems to accept these values. In addition, our schools are teaching race theory to promote hatred between the races instead of teaching the three Rs, history, and civics. In a new world we would have the opportunity to start over and instill the values our founders brought to create the greatest country in the world. We can do it again on this exoplanet."

"That is interesting and profound," commented Jennifer. "You must be a political conservative. How would you manage a crewmate who has a strong liberal or progressive bent?"

"I would use logic and reason to turn them around. I know in time I will succeed. I know that one's environment while growing up can mold one's political thinking, but intelligence and an open mind will win out in the end. The crew should eventually be of one political mind in order to create a harmonious colony."

"What about diversity?" questioned Owens.

"It's not clear that diversity has been a positive force in our democracy. When our country was founded and new immigrants were accepted into our society, they learned to speak English, accept the Constitution and the Bill of Rights, and put their parochial views aside to assimilate for the good of the Union. But emphasis on diversity has fractured the country into separate tribes that have little interest in the strength of unity. In a new world, we have the opportunity to bury self-serving interest for the good of the new community. Just think of how much we could accomplish if our politicians were not so polarized as they are now."

Changing the subject, Michael asked, "How do you intend to utilize the abundant free time on this long journey?"

"First, growing our food will be my top priority. But also, I am a lover of music. I intend to bring my guitar onboard to play a few hours each day, provided my crewmates enjoy it. I love to sing and am told I have a melodious voice. I would hope my crewmates would join in. I also expect to download an extensive playlist of my favorite songs on my laptop for us all to enjoy. Its sound system is exceptional, and I think it would be a great benefit for all of us to spend some of our downtime with music. Finally, I expect my crewmates will have activities in which they would Invite me to participate. I look forward to developing great friendships."

Sheryl, asked, "What is your sexual orientation?"

I am heterosexual and hope to get married and raise a family. So far, my dating has not been successful. I have been sexually attracted to a few men, but I have not found someone compatible with my strong political convictions."

Sheryl received the distinct impression that Samantha was hiding something but was unable to pinpoint it. She asked, "What is your ideal mate?"

"A strong masculine type who is sensitive and loving in addition to sharing my political views and my love of music." Samantha stole a look at Michael.

"How do you feel about leaving family and friends behind?" quizzed Tony.

"I have thought about that a great deal, and concluded the sacrifice is worth it to have the opportunity to be part of a new community and strive to create an ideal colony."

"Any questions for us, Samantha?" asked Tony.

"Oh yes! Big ones! What are the risks of never successfully reaching this exoplanet?"

Kevin responded, "We are dependent on the Starship's propulsion and navigation systems. The nav has been proven many times as have the Raptor chemical engines. Engineers at Los Alamos Laboratory are evaluating the nuclear ion engine as we speak. But its design is not simple. At this stage we are not sure about its exact performance, but we will know it well a year from now when we launch. The performance will tell us whether our estimate of an eight-year mission is accurate."

"What if you are wrong about the exoplanet's benign environment?"

"We selected this planet from 2,500 other Earth-size planets discovered by the Kepler telescope which could be promising places for life. Actually, based on extrapolating Kepler data, there could be as many as forty billion rocky Earth-size exoplanets orbiting in the habitable zone of Sun-like stars and red dwarfs in the Milky Way. But even if the atmosphere and temperature of this planet are not as ideal as we believe, we've proven how to deal with the challenging conditions in our lunar colony."

"This is exciting," Samantha responded. "What are the risks?"

"The risks of the unknown are always present, but we have learned enough to mitigate them."

Taylor interjected, "Now, if you have no other questions, Samantha, we will deliberate and call you soon with our decision."

When Samantha left the room, the committee offered their opinions. "Her skills are exactly what we need but will her strong political views hinder crew compatibility?" asked Taylor.

"Not for me," offered Kevin. "I agree with her views."

"Works for me if her vetting checks out," said Jennifer.

"Ditto," agreed Michael. "I really like her."

He added, "This is no simple farm girl. She knows her values. She strikes me as deep, intelligent, mature, and a real asset to our crew."

He neglected to add that he was physically attracted to her.

"Okay, let's conduct a thorough vetting with her references and colleagues with emphasis on her ability to be part of a team," declared Taylor.

Subsequently, she received high evaluations in every trait. They invited Samantha Gere to join the committee for their next interview.

***

Taylor introduced them to their next candidate, Daniel Montgomery, a 29-year-old strikingly handsome high school teacher from Los Angeles, California.

After his obligatory remarks and introducing Montgomery to Sheryl, Kevin, Jennifer, Michael, and Samantha, Tony began. "May we call you Dan? We want this to be informal and ask you to use our first names, rather than titles."

"Of course; please do," Dan responded.

"Dan, what do you bring to this team?" asked Tony.

"I am an excellent educator, experienced in teaching preschool, elementary, and high school. The colony will have children with the opportunity to develop and educate them properly. I am excited to succeed at this challenge."

"We need to ask you some very personal questions to determine your compatibility with the crew members with whom you will be spending the rest of your life."

"That makes a lot of sense to me. I welcome such a discussion."

"First, why did you volunteer for such an unusual way to spend the rest of your life?" asked Sheryl.

"In addition to educating our children, I can't think of a more rewarding way to spend the rest of my life. It is a chance to right the wrongs of our current society and start a whole new civilization with proper ideals for humanity."

"Please elaborate," said Kevin.

“Somehow our country has degraded into a socialist state. We elect socialist politicians, and it is killing this country. Too many Americans have decided to live off unemployment checks, food stamps, subsidized housing, and welfare. These are the people who re-elect the politicians that promise more free stuff. We have a serious shortage of real workers as increasingly able-bodied people decide to stay home and collect welfare checks. As Margaret Thatcher once said, ‘Eventually you run out of other people’s money.’ What ever happened to the American dream? The deficit spending and national debt created by our politicians is unsustainable. I don’t see any way out.

“Crime is rampant and getting worse every day. Instead of clamping down, we coddle criminals, reduce their sentences, defund the police, and treat the perpetrators as victims of society.

“America is no longer the leader of the free world. Our national leaders are incompetent, weak and corrupt. Someone once said that power corrupts, and absolute power corrupts absolutely. We have opened our borders to those who expect to live off the government. We are being infiltrated with terrorists who want to kill us. And our leaders don’t know what to do about it. They stick their heads in the sand, pretend everything is fine, and tell us to drink the Kool-Aid.

“A decade ago, we finally elected a president who implemented policies that put America first and we began to solve the problems. But the deep state impeached him and voted him out of office.

“Every year our country is becoming more divided. Instead of the UNITED states, our politicians and electorate pit one state against the other, vying for more power. After decades of real progress, we now use race to further divide us. Our schools are teaching Critical Race Theory to define victims and oppressors. Our schools continue to degrade, producing high school graduates who can't write and don't read. Grammar is a dying discipline. Even our news anchors and sports commentators don't know which pronouns to use as the object of a preposition. We no longer reward achievement and exceptionalism but give diplomas as participation awards. We don't reward our best teachers nor fire the worst tenured ones, even though everyone knows the good ones from the bad ones. To avoid hurting their self-esteem, we are telling our students that they are the best. We give them trophies for attendance, not for achievement. They grow up believing in the lie that they don't have to work hard to excel. If you ask high school graduates where they rank in any subject: mathematics, literature, history, science, English, they tell you they are the best. Yet they really rank below 25th in the world. We dumb down our grading system and curriculum to make sure we don't hurt their feelings or self-esteem. We are raising our young people to fail and assure them it is no fault of their own. America is doomed. These will be the inept future leaders of this country.

"In the words of Sammy Davis Jr., 'Stop the world, I want to get off!'

"So, for me, this is the only opportunity to wipe out these problems, start fresh and create a community of productive, industrious, educated individuals who are self-reliant. I want to play a vital role in reaching this goal."

Tony interjected, "Thank you for your candor, Dan. You've given us all much to consider. Any questions?"

"Of course, I do. I want to know whether any of our future colonists subscribe to the left, progressive, socialist agenda which has taken root in America these days. These people refuse to accept that what they are doing undermines the values of a productive society.

They will destroy any chance of success and sustainability."

Kevin responded, "I can tell you that I have political views similar to yours, but there are gray areas that are not so black and white. We intend to seek teammates with similar views to ensure crew and colony compatibility."

"Dan, what have you done personally to counter any of these problems which you obviously feel so strongly about?" asked Michael.

"I can't do anything about them. As a teacher they would ostracize me and maybe even fire me if I tried. That's why I am so frustrated."

"Please tell us about your sexual preferences," requested Sheryl.

"I am straight and sexually attracted to women. But so far have not met any who meet my standards."

After Dan left the building in a huff, still agitated, Tony asked for comments. Kevin offered, "I can't say I disagree with any of Dan's major points. But his views are too strong to be a good crew member. He is much too negative, too intolerant, and too emotionally vested in his strong, negative opinions. We need teammates with a positive attitude."

All seemed to agree. Dan was rejected.

Later that afternoon, Tony announced, "Our next candidate is Candice Li, a 23-year-old physical therapist from Houston, Texas."

She made an impressive entrance with jet black hair cut short, wearing a short, tight skirt, crisp blouse, and a winning smile. She had exceptionally beautiful legs and long eyelashes.

After Tony made the usual introductions and soliloquy, he began, "Candice, please tell us about yourself."

She smiled at Michael and Kevin, obviously flirting.

She responded to Tony, “I have been a yoga instructor and physical therapist since I graduated from the University of Texas in Austin. But my real passion is space. I received a minor in space studies from UT. I applied for astronaut training last year but was not accepted into the program.

“My training in PT and yoga should be valuable for the long flight as well as in the colony.

“My parents immigrated to the United States from China before I was born. They returned to China in 2020 to join the resistance to the takeover of Hong Kong. The Chinese Communist Party arrested them, and I have not heard from them since. I tried to get the US State Department to secure their release, but we have made no progress. I’m afraid all is lost. I have no knowledge as to whether they are still alive.

“I have been a solid Republican since that time. However, the Republicans’ lack of progress in solving the problems that eat away at our liberties is frustrating. Maybe I should be a Libertarian.”

“What is it about the Republican Party that aligns you with it?” asked Tony.

“They have adopted 14 principles that resonate with my thinking.”

“Please summarize them for us.”

“Small, fiscally sound federal government; free enterprise; individual self-reliance.”

“Why are you interested in this mission?” asked Owens.

“At first, it was my friends that discerned how much this mission would mean to me. As I learned about it, I realized they were right. Lately I have become obsessed with the idea of forming a new country free from the problems facing our country today.”

“What is your sexual orientation, Candice?” asked Sheryl.

"I am straight, a healthy heterosexual, very interested in getting married and raising a large family. I dated in college and somewhat since that time but haven't found the right man as yet. My friends keep trying to fix me up with several men, including other Asians, but I've learned most do not truly believe in equality of the sexes. I am my own person and resist the idea that women should be subordinate to a man. I have high self-esteem and believe I can accomplish anything I wish. I look for a man who accepts me as his equal. Kevin and Michael, please tell me whether you can accept this."

"Of course," responded Kevin. Michael enthusiastically nodded his agreement. Jennifer was pleased with their strong assurances and not at all surprised.

"That is a relief. I don't mean to be so forward, but I assume the idea is to colonize the new planet with the six of us. And that means we will be expected to pair up and procreate."

"You are correct, Candice," assured Taylor.

"What qualities would make a man appealing to you?" asked Sheryl.

"Physically, I'd prefer he be trim and good looking. I think he should be even-tempered, intelligent, and have a sense of humor. But most of all, it's important that he share my values."

Sheryl followed up, "Are you religious?"

"Yes, I have adopted Christianity. But I haven't been a regular Church attendee. I've tried several churches but haven't found one I can call home."

Candice asked, "I notice that thus far I would be the only non-Caucasian. Is that by design?"

"Not at all," responded Tony. "Our selection is based solely on the qualifications we need for the mission and crew compatibility. We do require that all crew members be fertile heterosexuals, and that the women are of childbearing age at the time of exoplanet arrival. Ethnicity is irrelevant, but compatibility is paramount."

"How would you intend to spend so much time en route to the new world?" asked Kevin.

"I have always been interested in writing but have not published anything. I would love to keep a journal of our odyssey. I also expect to devote time to teaching and coaching the crew how to enhance and maintain their bodies."

"What do you believe are the challenges facing our country today?" asked Michael.

Candice responded, "They are mostly foreign policy issues beyond those I have already discussed. But one big issue is our shrinking population growth. But I don't know how to solve this. Back to foreign policy, as long as our politicians are so polarized, I don't see how we can resolve these either. I believe the Communist Chinese Party does intend to dominate the world, just as they have declared publicly. We could stop this if we got our act together and stood up to their theft of our technology, infiltrating our universities and companies, indoctrinating our students with socialism and communism, and flexing their military muscle in Hong Kong and the South China Sea. Our politicians just scold them and let it happen. Next step is likely the invasion of Taiwan. It hurts me to helplessly watch this happen. I have expressed my concern to the administration but haven't even received a response. I doubt they have the fortitude to stop any invasion. In my opinion, they fear China, which is manifested in their reluctance to do anything about standing up to the CCP aggression.

"I am also appalled at the escalation of crime in our cities. I lay this at the feet of the local politicians who want to defund the police, reduce the size of the police force, reduce the bail and sentencing of lawbreakers, and minimize charges against these thugs. And I don't see the voters extracting these misguided politicians from office."

"What if one of your crew members does not agree with your concerns?" asked Sheryl.

" I would want to talk with him before I sign on. If I could not convince him of my position, I would decline an invitation to join the crew. Disagreements on these important topics would not be a good basis for forming a new world colony. I feel very strongly about this."

"Thank you for your candor, Candice," said Tony. "We agree that the crew should not have strong, diametrically opposed views. Not just for forming a colony, but for traveling together for eight years."

"Do you have any other questions for us?" asked Tony.

"Yes. Please provide feedback for me on the political views of the crew."

Kevin answered, "Based on the interviews to date, I am sure we are all politically compatible. We all share your conservative views and concerns about our country. Do you all agree?" All three nodded emphatically.

Tony declared, "If you have no more questions, we will deliberate and get back to you soon."

When Candice left the room, Kevin said, "Whew! She is a breath of fresh air, not holding anything back. A strong woman. I like her!"

"What do the rest of you think?" asked Tony.

"I totally agree," said both Samantha and Michael," emphatically.

"I don't disagree," said Jennifer. "But once again we have to execute a thorough vetting to ensure her strong positions will not cause any significant issues en route nor in the colony."

The vetting completed; Tony called Candice to invite her to the crew. He added that her strong position on issues were compatible with her crewmates. He added that she was free to explore this with each personally. He also invited her to join the committee to help select the final crew member.

***

Taylor convened the committee once more. "Okay, we have one more crew member to choose before we begin to select a backup crew."

They gathered to meet Jason Appleton, a handsome 27-year-old computer scientist from Boston, Massachusetts with a strong hobby in astronomy. After Michael's introductory remarks, the committee introduced themselves and began their questions. Candice again liked what she saw. Jason was equally captivated with the appearance of the female crew members.

Taylor began, "Jason, please tell us about your qualifications for this unusual mission."

Appleton began, "My degrees and career in software development and artificial intelligence should be valuable in programming the advanced computers I'm sure we will have onboard and bring to the new world. And I understand we will have a humanoid robot onboard. I hope I will be able to use my artificial intelligence background to program it. The needs we will encounter in both venues are not even known at this time but could be limitless. In addition, my background and knowledge of astronomy may be valuable in celestial navigation and exploration. This mission is more than I would ever hope to experience in my lifetime and my talents should be vital to its success."

"You are not married and have no children but are you sure you will be able to leave your relatives and friends?" asked Taylor.

"Well, I've obviously carefully considered this issue, but I would hope to make strong friends with my crewmates and develop an intimate relationship with a female member. I am a healthy heterosexual, but to date have not met a woman who shares my values so necessary for a long-term relationship."

"What are those values?" asked Candice.

"I am a registered Independent politically. I believe strongly in self-reliance. I reject the growing social programs that enable the government to subsidize people who do not work for a living, Programs like lucrative unemployment stipends, childcare subsidies, free housing, and food stamps make too many citizens dependent on the government. And their multiple generations of dependency are perpetuated indefinitely. The world does not need this societal drag on productivity. Our politicians think we can borrow and spend indefinity. We now have a 33-trillion-dollar debt, which likely cannot be repaid. Our corrupt politicians added four hundred billion dollars to this debt by forgiving student loans to buy votes. All that did was screw the students who paid off their loans. Moreover, people who didn't go to college end up paying for those who did. And this just encouraged greedy colleges to increase their exorbitant tuition. China owns much of this debt. And they will demand their pound of flesh eventually.

"This nation was founded on self-sufficiency, and as a result, we grew and prospered. I am concerned that we are losing these values. Getting an education and working hard benefits us all. I am seeking a spouse that shares these views.

"In addition, I am a religious man and believe my spouse needs to share my religious beliefs to have a happy marriage."

"What is your religion?" asked Samantha.

"I have been a practicing Christian for most of my life," answered Jason.

"What do you hope to achieve on this one-way mission?" asked Kevin.

"As I stated earlier, I hope to find the love of my life and make lifelong friends. Of course, this can be done here on Earth. But I also hope to fulfill my dream of exploring the Solar System and beyond. Colonizing a new planet and creating a new, perhaps perfect, society, applying the lessons learned from Earth and rectifying these problems would be the ultimate achievement."

"Any hobbies besides astronomy?" asked Candice.

"Well, as a teen I got deep into hypnosis. I read extensively about the field. I was fascinated by the eighteenth-century German physician, Franz Mesmer. His hobby was also astronomy, making him a kindred spirit. He accomplished extraordinary success with what he called animal magnetism, later called mesmerism, and then hypnosis. He incorporated it into his medical practice. I also studied the antics of the stage hypnotists and wondered about their power over subjects from their audience. When I was 17 years old, I was a smoker and wanted to quit but failed miserably. I asked a friend, who was an accomplished hypnotist, if he could help me quit. He easily put me under and told me smoking a cigarette would taste like seaweed. It worked! I haven't smoked since. Hypnosis became an entertaining and useful hobby after I learned the technique and how to avoid the detrimental effects, which can be crucial."

"Fascinating," said Candice. "I always thought hypnosis was a hoax."

"Do you have any questions of us?" asked Taylor.

"Several. How long is the training program before we leave Earth?" Jason requested.

Taylor responded, "We are planning for twelve months, but that will depend on your progress. You will have both classroom training on the various systems, including the new state-of-the-art flight computer and its software. You will spend significant training time in the spacecraft simulators. Although each team member has a specialty, all will learn each other's disciplines and duties to some extent, introducing a degree of fail-safe redundancy. You will also be trained in colonization to the extent that we can hypothesize environmental conditions on the exoplanet."

Jason followed up, "Will NASA pay our salary during the training period?"

"Of course."

Jason continued, "Will we have any transportation aid to enable planet exploration?"

Kevin answered, "We are modifying the rover used on the lunar colony to adapt to the presumed regolith texture of the planet's soil. We will store this in one of the many Starship bays. I drove one on the Moon and it worked great!"

Jason asked, "Will we explore the planets and their moons before we leave the Solar System?"

"Somewhat," Tony replied. "But you will be traveling at high rates of speed, so your observations will be time-limited and distant. Our primary objective is to colonize the exoplanet. But if the journey does not deviate from this objective, there may be some opportunity to see portions of the Solar System.

"Jason, If you have no further questions at this time, we will be contacting you in a few days," declared Tony. "Thank you."

After Jason left, Taylor polled the committee members and received a "thumbs up" from all.

After another thorough vetting, Taylor called Jason and received an enthusiastic acceptance.

"When can I start?"

The NASA committee members alone began selecting the backup crew without the primary crew members who were busy beginning their training.

4. Preparation

The crew welcomed Jason to the team. They all were excited to begin their training and mutually impressed with their crewmates' backgrounds and appearances.

Kevin decided the first order of business was to assemble the team to get to know each other better before they started training. He called a meeting at a JSC conference room. Everyone was anxious to start.

He began, "Thanks for coming. I thought we should get to know each other before diving into the strenuous training schedule. We all come from diverse backgrounds and disciplines, and perhaps have varied reasons for volunteering for this unique mission. You have all read each other's biographies, but let's share something about ourselves that will begin to help us understand each other and thus may begin to unify us as a team.

"First, I want you to know how happy I am to serve as your mission commander. We have assembled an excellent crew. I am very impressed. Each of you has an outstanding record which will contribute to our success. I hope in these next twelve months we will know each other personally as well as professionally. For me, exploring space has been my most important goal ever for as long as I can remember. And to undertake this important, thrilling mission reaches the pinnacle of my aspirations.

"The training will be rigorous, but I know each of you will appreciate the results. If anyone believes it falls short of your needs, let me know and we will delay the mission until we are ready. But we can't delay the launch for long because we want to take advantage of the alignment of the planets to accelerate our Starship beyond that provided by our propulsions system.

"We will go beyond the technical training to include psychological evaluation and preparation. If you have reservations about this tremendous commitment and wish to return to your family and friends, I will understand. But let us know early, so we can replace you with a member of the backup crew.

"Any fame we experience will be short-lived as we will not be returning to Earth to enjoy it like the Apollo astronauts did. So, I hope your one year of fame is not part of your motivation for the mission."

The team traded information on their respective backgrounds, why they had volunteered and what they expected from the mission. The bonding had begun.

"Candice, I loved your interview," offered Kevin. "You said you were a Republican. What are those fourteen principles that impressed you?"

"Sure. I have them committed to memory. They make up an important part of my convictions:

1. A belief in smaller government.
2. Support of a federalized system of government to bring power closer to the people.
3. Fiscal conservatism with limited taxes and constrained government spending.
4. Strong national defense.
5. Individual liberties and responsibilities.
6. Tolerance, inclusiveness, and optimism with the right to disagree and debate on any issue.
7. Government actions are limited to issues which cannot be done by individuals.
8. Government is based on the individual, honoring ability, dignity, freedom, and responsibility.
9. Free enterprise, encouraging individual initiative and incentive.
10. Sound monetary policy.
11. Equal rights, justice and opportunity independent of race, creed, age, sex, or national origin.
12. Retain important principles yet be receptive to new ideas and varying points of view.
13. Value national strength and pride with desire for peace and freedom throughout the world.
14. Incorporate these principles into principles for government."

"Wow! That is powerful," responded Kevin.

"Works for me," added Michael.

"Me too," agreed Jennifer.

"I'm not a Republican, but a registered Independent," volunteered Jason. "But I don't disagree with any of these."

"Well, based on these, contrary to my younger days I guess I must have always been a Republican all along," added Michael. "But don't Democrats also adopt these principles?"

"I think they used to," answered Candice. "But today the party has been taken over to bastardize these tenets. Based on their recent legislation, they certainly don't believe in smaller government, constrained government spending, nor government actions limited to issues which cannot be done by individuals. Moreover, too many politicians of all stripes, Republicans and Independents included, just give lip service to giving power to the people, individual liberties, and equal opportunity. Just look at the mandates of Affirmative Action, wherein preference is given to race rather than achievement or merit."

"Okay," laughed Samantha, "since we're so compatible, let's all gather into a circle, hold hands, and sing *Kumbaya.*"

***

NASA put them up in a large house in El Lago, Texas four miles from the Johnson Space Center off NASA Road 1. Kevin was pleased it was the same town he had chosen when he first began at JSC. The small town of 3,000 is considered the "Home of the Astronauts," because so many lived there in the early days of manned spaceflight. NASA had modified the three-bedroom house into six tiny bedrooms, tailored to simulate their sleep pods in the Starship spacecraft.

Their individuality was limited externally to the cars they leased. Kevin and Candice drove Teslas, appropriate for the company's tie to Elon Musk's SpaceX launch vehicle and Dagon capsule.

The classroom training was conducted at the Johnson Space Center. In addition, the crew conducted much of the simulation training there. The simulation consisted of a full mockup of the new SpaceX second stage, the Spaceship. It consisted of the ten decks of living space and storage plus the decks holding the engines and fuel. The Dragon capsule used for training was a perfect replica with the spacecraft electronics, displays, and actuators. They learned to use an exact copy of their new, powerful onboard computer, including the software that served to control the engines and navigation. The trainer also simulated the cupola which they used for sighting the environment outside the spacecraft. Only the training for the launch portion of the mission took place at Cape Canaveral under the watchful eye of the launch mission controllers at the Cape.

Kevin was disappointed that he would not be flying the NASA T-38 as often as he did for the lunar mission. But the other crew members, who had no pilot training, flew commercial. When Kevin did fly to the Cape, one of his crew members flew with him since the T-38 cockpit held two people. In five flights to the Cape, each of his crew had their turn, while Owens showed off his expertise with a few aerial maneuvers for their exhilaration. He especially played "Top Gun" for Jennifer. On the way to the Cape, he flew supersonically over the Gulf of Mexico and broke a couple of FAA rules doing barrel rolls, flips, and inverted maneuvers. But instead of getting sick, Jennifer loved it. She planted a wet kiss on his lips when he helped her out of the cockpit. He happily kissed her back.

They all stayed at the historically traditional Holiday Inn in Cocoa Beach for each trip to the Cape.

Their training included learning hand signals and gestures to communicate in the event they lost their communication system either on the spacecraft or at the colony, much like scuba divers and military pilots do. Their facial expressions also conveyed vital information, such as quizzical frowning, saying, "I don't understand" or "I don't agree" or smiling, conveying "I'm with you on that."

Each Sunday Michael and Jason attend the Presbyterian church service in Webster, just down the road on NASA Road 1 three miles west of JSC. Kevin usually went with them. Occasionally all six attended.

After eight weeks, NASA's Johnson Space Center psychologist called each team member in to do an evaluation by rating their peers to assess their likelihood for mission success. They were told this would be in complete confidence and not to disclose this evaluation process or even that they participated in the process. The process was conducted by the JSC Astronaut Office based on a rating system developed by Robert Rose and R. L. Helmreich in 1983.

Five traits were evaluated.

Job Competence:

- Absorbs information rapidly.
- Reduces complex issues into solvable partitions.
- Accomplishes tasks thoroughly and efficiently.
- Develops innovative solutions to difficult problems.
- Consistent performer under stress.
- Manages mental and physical stress.
- Avoids unnecessary risk.

Communications:

- Presents himself/herself well.
- Speaks clearly and effectively.

Leadership:

- Motivates others to complete tasks.
- Is not authoritarian.
- Is decisive when required; flexible when needed.
- I would follow his/her lead.

Teamwork:

- Puts group goals ahead of individual.
- Works effectively with every member of the crew.
- Pulls his/her own weight.
- Shares credit; accepts blame.

Personal/Group Living:

- Easy to get along with; comfortable.
- Good listener.
- Absence of irritating qualities.
- Considerate of others.
- Wears well over time.
- Helpful.
- Tolerant of individual differences.
- Has a sense of humor.

Although most resented evaluating their peers, they acquiesced. NASA was overjoyed with the results. In the words of Author Tom Wolfe, they had "the right stuff." Each crew member passed with high evaluations. Only Owens was cited by some for being authoritarian. But this was expected for the mission commander. They emphasized that they accepted and respected Owens' authority and leadership as well as intensely liking him as an individual.

The media was given a press release including a brief bio on each. Although they did not receive the acclamation of the original Mercury 7 astronauts, it came close. The mission was extraordinary: six Americans would leave the Solar System and attempt to explore and colonize a strange new planet four light years away! The media was just as interested in the backgrounds of the crew as they were in the mission. Most believed the mission would fail. So, they emphasized their coverage on the makeup of these brave civilian astronauts and their motivations as much as the adventure facing them.

Kevin, being the only military astronaut and the only crew member who had been to space and the Moon, naturally received the most attention. They wanted to know his life history. He was offered a book deal but declined. He did not have time for the distractions.

For Owens, much of the training was a repeat of his training for the lunar mission, but the other five had more difficulty because everything was new. The first flight with the "vomit comet" resulted in nausea for three of them. But eventually they all withstood the weightlessness experience, and some enjoyed it.

They also had fun in the JSC Neutral Buoyancy Pool, but not so much in the centrifuge. This assessed their ability to function under the force of more than 4 G's, which they would experience during the SpaceX launch. This was similar to the 3 G's experienced during Apollo, Skylab, and Shuttle launches. They all excelled in the classroom training, exhibiting their intellect and desire.

Jennifer kept looking longingly at Kevin. *He is such a hunk,* she thought. *I know he is interested in me. But why doesn't he make a move? I guess I'll have to nudge him a bit.*

Kevin and Jennifer did a little extra bonding in the pool. Jennifer brushed her breasts against Kevin's arm. He thought it was an accident until he saw her wicked smile. He returned the favor with a squeeze of her cute derriere as she was getting out of the pool.

She turned to him and said, "Kev, don't start something you can't finish."

Winking at her, he said, "Hey, Jen. You started it. But trust me; I CAN finish it."

Smiling, she retorted, "Promises, promises."

"I think you need a slow, soft, erotic spanking, young lady!"

Winking back, "Promises, promises."

Hmmm, thought Jen, smiling. *I guess he's not so shy after all.*

This exchange did not go unnoticed by the rest of the crew. They were all looking forward to their "special bonding."

That night, if you listened closely, giggling, heavy breathing, and soft moans came from Dr. Thorson's room.

Kevin made it a point to continue the crew bonding after hours. After training at JSC, all six gathered for drinks at a favorite watering hole in Seabrook. Then they went for dinner at either the Volcano Room on NASA Road 1 or Playa Maya Tacos and Grill on Highway 146 in Kemah. They especially enjoyed the fresh seafood and oysters at the Volcano Room and the Tex-Mex at Playa Maya. *We will not get this great food in space*, they lamented.

They all attended the Seabrook watering hole on a special karaoke night. Sam loved it and couldn't contain her enthusiasm. She grabbed Mike's hand and exclaimed, "Let's do it!"

"Do what? Sing?" asked Mike. "What would we sing?"

"I have an idea," she smiled as she dragged Mike up to the stage. She selected the song and began: "Fly me to the Moon." Mike joined her. "Let me play among the stars. Let me see what spring is like on Jupiter and Mars."

The rest of the team beamed attentively as their eyes lit up. When Sam and Mike finished their number, Candice jumped up, cheering and applauding. Jason, Jen and Kevin soon followed suit. Eventually the whole crowd gave a standing ovation. "Encore!" they shouted.

But Mike was embarrassed, so Sam took his hand, and led him off stage to join their teammates.

Candy exclaimed, "You guys are REALLY good! Sam, do you sing professionally?"

"No, I just have fun singing," laughed Samantha.

"We ought to bottle that rendition and take it with us," added Kevin. "We could adopt it as our theme song."

That night Sam and Mike celebrated their own special bonding in Sam's room.

The next day, the team met to solidify their program of cross training suggested by NASA to enable backup for their roles both en route and in colony life. Michael was overjoyed with being selected as Kevin's backup commander. Kevin agreed to learn engineering from Mike, concentrating on Starship's systems. Candice was eager to expand her limited medical training with Jennifer to include midwifery. Jennifer was happy to study agronomy from Samantha. Sam was interested in what Jason could teach her about computer science. Jason was equally intrigued about Candice's instruction in physical therapy. They would all have ample time to solidify their learning of their secondary disciplines during their eight-year voyage.

Much of their training was in the Starship simulator. Most were amazed at the enormity of the spacecraft. SpaceX had elongated the original Starship second stage by another seventy-two feet. At the apex of the 232-foot-tall cylinder was the observation cupola into which all six could squeeze for a 360-degree view of space. It was covered with a shroud to protect it during launch and atmospheric entry. Directly beneath the cupola were the astronaut couches, enabling them to withstand the launch and entry G forces while lying on their backs. Each deck provided a hatch to traverse between the top ten decks. Directly forward and overhead in the top deck were the new Dragon touchscreen command and control panels. The third deck housed a circular central table with six chairs, all mylar blow-ups fastened to the deck's floor. They were to be used for meals, meetings, and recreation. It also held their three thousand pre-packaged, freeze-dried, dehydrated meals stored in shelves around the perimeter. In the center of Deck 4 covering the floor was the hydroponic garden for growing fresh vegetables and fruit. Deck 5 housed their gym and exercise equipment. In Deck 6 were their sleeping pods arranged in a circle around the center with lightweight partitions between the pods for privacy. A shelf for clothes was provided for each. Deck 7 contained the shower quarters and their water supply. The eighth deck housed the oxygenator, life support equipment, pharmaceuticals, and medical apparatus including an ultrasound machine and x-ray machine. Deck 9 housed the spacecraft supercomputer, their personal computers, iPads, the two 3D printers with the Fusion 360 software platform for CAD/CAM modeling, and a few of Kevin's tools. Deck 10 was used for storage including the folded rover vehicle, a microscope, an imaging spectrometer, several tools, and the advanced humanoid robot, originally developed by Tesla in 2023. Decks 11-16 housed the engines, liquid xenon fuel for the nuclear electrical propulsion engine and liquid methane fuel and liquid oxygen for the Raptor engines. Fuel lines also fed the external attitude control jets. The crew joked that the spacecraft needed a center pole for pole dancing entertainment.

Kevin convened the team at JSC. "NASA wants us to design our own logo as was done for every NASA crewed mission. A patch that signifies our mission but also identifies us."

"Oh, I was hoping we'd get to do that," said Jennifer. "I collected all the Apollo patches. I love the ingenuity they used for each mission. I was worried that since five of us are civilians, rather than astronauts, we wouldn't have this opportunity. Will our names be on the logo?"

"That is up to us, but every Apollo crew except Apollo 11 and 13 did so.

"So, give me your ideas."

"I think it should show us landing on Alpha Proxima b," said Candice.

"How about showing our rendezvous with the lunar Gateway?" suggested Michael.

"I think we should show the majestic liftoff of the largest rocket ever," said Samantha.

"Or, what about showing the Solar System behind us as we speed away?" offered Jason.

"But can't we show the Alpha Centauri triad as we approach it?" asked Jennifer.

"I like all those ideas," said Kevin. "You guys have obviously given this some thought. NASA is assigning a couple of graphic designers to us. Let's lay these thoughts on them to see what sketches they can produce, then get back together and agree on the best one."

One week later they met with the artists, excitedly scrutinized the sketches, and solicited their opinions. They selected a sketch of the Starship accelerating away from the Solar System's planets. This scene was on a black background with bright stars. The names Owens, Thorson, Stone, Li, Appleton, and Gere surrounded the scene in a circle.

They were all overjoyed with pride as the sketch was made into a textile patch and embroidered into their jumpers and spacesuits. It was also released to the public for purchase as souvenirs.

***

NASA arranged a press conference for the "Special Six," as the media dubbed them. The conference room was jammed with newspaper reporters, all the TV networks, video equipment, and microphones. The Special Six took their places sitting across the dais, proudly wearing their new jumpsuits and mission patch. Kevin's also displayed his NASA astronaut patch. Each crew member had their name prominently displayed in front of him/her with their new logo depicted on their nameplate. This was very reminiscent of the first press conference held 70 years earlier for the seven original astronauts.

Some of the questions were inciteful, others not so much. Answers tended to be brief, but complicated topics required more elaboration.

"You six are the most famous people on the planet right now. More famous than Neil Armstrong. How do you feel about that, Commander?"

"Fame doesn't matter that much to any of us. In a couple of months, we will be leaving Earth forever. So why should we value fame?"

But most of them thought, *Hey, this is my fifteen minutes of fame. Let me enjoy it.*

"We have been told that your objective is in the Alpha Centauri triumvirate of stars. Please tell us about your destination."

"Alpha Centauri looks like one of the brightest stars from Earth. But it really is three stars more than four light years away, which are mutually attracted gravitationally. Alpha Centauri A is the largest and brightest of the three; Alpha Centauri B is smaller and one hundred million miles from A. Both are M-type stars, like our Sun. They both revolve around a common barycenter with a period of nearly 80 years. Alpha Centauri C, also called Proxima Centauri, is the Sun of our destination planet. Proxima Centauri is a red dwarf of spectral class M6-Ve. It is 20,000 times fainter than the Sun, which is why it is not visible from Earth. We are landing on a planet in its habitable zone, called Proxima Centauri b."

"Do all three stars have planets?"

"We don't know for certain. Proxima Centauri has three that we are aware of, but only our chosen planet is in the habitable zone. We think we found a gas giant around Alpha Centauri A. We also believe we detected an Earth-size planet around Alpha Centauri B. There may be others. Detecting planets outside our Solar System is difficult."

"Does your destination planet have water and land like Earth?"

"We really don't know. We know it is in the habitable zone and therefore we think it has liquid water, but we don't know the percentage of water to land. Astronomers believe that only one percent of habitable planets have two-thirds water and one-third land like Earth does."

"Commander Owens, I heard you turned down a lucrative book deal. Why are you undertaking such a dangerous mission, when you could remain on Earth and make a fortune?"

"This is the adventure of a lifetime (no pun intended), and money is not important to me," answered Kevin.

"We can't help but notice how exceptionally attractive all six of you are. Was that a consideration in crew selection?"

"Well, to some extent, yes," answered Kevin, embarrassed. "Don't forget, we are expected to raise large families to grow the colony for sustainability."

Also, don't forget we all chose each other, thought Samantha, smiling.

"So, you were tested for fertility?"

"Absolutely."

"Commander, you are five Caucasians and one Asian. Did you not emphasize diversity, consistent with the governmental woke culture and inclusion objectives in your selection process?"

"No. We reject such objectives. We emphasize compatibility to maximize crew unity."

"You all will be crammed together in a steel tube for at least eight years. Won't you get tired of each other and get on each other's nerves?"

"We recognize the potential for human strife in such conditions, so we did all we could to ensure crew common values and strong opinions and beliefs. We have spent almost one year living together and found our common values to be extraordinarily compatible. As a result, we found a strong bond, mutual respect, and unity. We are confident this will pay dividends through-out our mission."

"What are those common values and opinions?"

"We are all politically conservative and have some similarity in religious beliefs."

"So, none of the crew is liberal or progressive?"

"No. That degree of diversity would create compatibility problems, both during our long voyage and especially in establishing a unified colony."

"Wow. That could be interpreted as racism."

"That is absurd and insulting."

"Why has NASA restricted this to be an American mission, rather than international as has been successful involving several partners like the ISS and Artemis programs?"

"Same answer as stated earlier," said Kevin, bristling at the redundancy of the question. "But I will repeat it for your benefit. Please take notes on this because it is critical for mission success, and I don't want to answer it again. Diversity can cause conflicts that could jeopardize the mission. We have been diligent during our selection process to ensure compatibility and mutual respect between our crewmates. This goal would be difficult to achieve if we had competing values and diverse national backgrounds, histories, and priorities. We have learned from the International Space Station and the Artemis program that such diversity created unnecessary complications of crew incompat-ibility. Too often it created an 'US verses THEM' environment."

"Dr. Thorson, do you have any reservations about undertaking such a mission?" "Of course," Jennifer retorted. "But they are easily dispelled with this fantastic and unique opportunity."

"What if a crew member comes down with a serious medical problem of some kind?"

"We are prepared to handle almost anything. In addition to the physical examination and thorough blood test we all received prior to launch, I will repeat the blood tests on each of us annually. We have the equipment onboard to analyze the results and diagnose any anomaly."

"What if you encounter a medical emergency?"

"NASA has equipped me with all the pharmaceut-icals I requested, anesthesia, sterile paraphernalia, IVs, ultrasound and X-ray equipment, basically everything I need to deal with any medical emergency. Candice is trained to assist me where necessary."

"Jason, what do you expect to find on the exoplanet if you make it there?"

Appleton explained, "From what we know so far, there are similarities with Earth. So, the probability that its environment produced life in some form is high. We are excited about the prospect of discovering it."

"What is the likelihood that you find intelligent life?"

"No one knows, of course. But it is unlikely that only Earth has intelligent life. There are at least one hundred billion planets in the Milky Way galaxy.

Astrophysicists believe that three hundred million of these have the right ingredients to support some kind of life. Between 6 and 40 million are in the habitable zone of their star enabling liquid water to exist. Other considerations are whether the planets are rocky or gaseous, the characteristics of their star, atmosphere if any, surface temperature, and how long they have been in existence. As we know from Earth, life-form evolution is paramount. But this is only within our galaxy. There are at least two trillion galaxies in the Universe. So, what do you think is the answer to other intelligent life somewhere?"

"Will you have weapons to defend yourselves if you are threatened by some life form?"

"No," answered Kevin. "We have no reason to expect to encounter such a threat."

"What about hostile aliens from another galaxy?"

"You've been reading too much science fiction," laughed Michael.

"Commander Owens, how close to the speed of light do you expect to reach? What will happen to you as you reach maximum speed?"

Kevin responded," Good question. As yet, we really don't know precisely how fast the nuclear electrical propulsion engine will accelerate us. But we may reach speeds approaching 90% of the speed of light. In that case there will be some significant increase in our mass and dilation of time and distance as prescribed in Einstein's special theory of relativity. We will not detect the effects as they are relative to an outside observer. The mass increase, however, will reduce the acceleration produced by the nuclear fusion engine. If we reach 90% the speed of light, the mass of the crew, the spaceship, and all its contents will double, reducing our engine acceleration by 50%. But since our time dilation will also be 50%, our travel time to our destination will not change."

"I'm sorry I asked, since I don't have a clue to understanding what you said."

"But how does the nuclear engine work?"

"We will still use our traditional chemical high impulse-thrust rocket requiring massive fuel only for launch, high acceleration periods, and landing on the exoplanet. Our plasma rocket provides us with a low, continuous thrust as a direct fusion drive, using minimal fuel. It converts thermal energy from a small nuclear reactor into continuous electrical energy. The electrical fields heat and accelerate a propellant, forming a plasma. Then magnetic fields direct the plasma in the proper direction as it is ejected from the engine's nozzle. This will be used for the duration of interplanetary and interstellar propulsion. This nuclear fusion engine can supply power for years, even after we land on our exoplanet. We will also use this electrical power to maintain our life support, heat, and communication systems."

"Candice, what do you say to people who claim this mission is impossible?"

"Nothing is impossible," quipped Li. "I have faith in NASA and our crew."

"Dr. Thorson, how can your team survive such a long trip without going insane?"

"We will be busy performing unique duties, cementing our mutual relationships, observing new phenomena, and exercising our personal hobbies in our spare time. We expect to have plenty of distractions to occupy our time. Also, NASA has learned that spaceflight can have positive effects on mental health, provided that the crew is compatible. That is why we have ensured that we all have common values, thereby minimizing significant conflict between crew members."

"Why did NASA choose SpaceX for this momentous and dangerous mission, rather than wait for the development of NASA's own Space Launch System?"

"Well, you can blame that on the visionary Elon Musk. He vowed decades ago that humans should be a multi-planet humanity. So, he invested his own wealth to accomplish it. As a result, under his leadership, with NASA encouragement and support, his SpaceX system proved to be superior and cheaper for the Artemis Program and now for our mission than SLS. There was no need to wait."

"Commander Owens, do you have enough fuel to make it?"

"Yes. We intend to use the Deep Space Gateway in lunar orbit to top off the fuel and oxidizer tanks for the Raptor chemical engines for midcourse correction and landing."

"Why are the engines called *Raptors*? Were they named after the historical predator dinosaurs in the *Jurassic Park* movie?"

"No. Elon Musk named them after the magnificent birds of prey of exceptional flight capabilities that fly majestically over the launch pad, such as eagles.

"To amplify the answer to your question on fuel sufficiency, our refueling at Gateway will supply more than sufficient fuel. We will need to restart our chemical engines for course correction. Our nominal trajectory will be perturbed to take advantage of planetary gravity assists, then used to return to the desired continuation trajectory. The chemical engines will be used for landing on the exoplanet. We will use xenon gas as fuel for the nuclear engine. We plan to have 18 years' worth of xenon onboard. So, we will have plenty even if we need to return, which is not in our plans."

"How about water, food, and oxygen?"

"NASA has provided Starship with devices to generate and convert water and oxygen along the way. We will also top off our supply of water at Gateway. Onboard will be an improved version of the Water Recyclable System that has worked so well on the International Space Station. It also feeds the Station's oxygen generator. It will do the same onboard our Starship. The exoplanet is believed to have an abundance of water."

Samantha took over. "Our supplies include several months of prepackaged meals. In addition, my training specialized in growing vegetables onboard using hydroponics. I will be using a version of the Vegetable Production System used on the International Space Station. I expect to grow Chile peppers, radishes, Mizuna mustard, Chinese cabbage, red Russian kale, Waldman's green lettuce, and other vegetables. Also, we hope to grow Outrageous Red Romaine lettuce, and Space Lettuce that contains PTH which should prevent bone loss. I also hope to grow a few simple fruits. I will be using a bank of light emitting diodes to enable the plant growth. We will need these fresh fruits and vegetables to supplement and ultimately replace the prepackaged meals."

"So, when the packaged meals are exhausted, you all will become vegans. Dr. Thorson, won't that weaken your bodies?"

"Jen, let me answer that," said Samantha.

"On the contrary, we will improve our health by eating only plants. Our immune systems will improve, reducing the risk of disease. In addition, veganism will improve our minds, avoiding overabundance of meat proteins in our brains. According to the American Dietetic Association, humans have no nutritional or biological need for animal products.

"Do you realize that if everyone on the planet ate only fruit and vegetables, the planet would reduce energy costs, conserve space, reduce toxic waste, reduce air pollution from carbon dioxide, methane, and nitrous oxide, reduce long-term medical costs, and preserve water and food for human consumption?"

"What is your role, Candice?"

"I will work with each crew member to induce a regiment of muscular toning. It is easy to experience muscle cramps on such a long journey and I know how to minimize their effects. In addition, Dr. Thorson is teaching me to be a physician's assistant, particularly during delivery of babies."

"Dr. Thorson, what happens to your immune system on such a long flight? Astronauts aboard the International Space Station for long durations have shown susceptibility to colds and viruses."

"It is true that we could have reduced immunity to bacteria and viruses if we were in a microgravity environment for years. But the continuous thrust of the ion engine enables our bodies and our genes to thrive in a meaningful, but low gravitational environment."

"What other provisions will be onboard?"

"We will have a diverse supply of medicines and pharmaceuticals," said Jennifer.

"We also have a 3D printer with construction material and additives," stated Michael.

"Also, clothes, cloth, and paraphernalia to make new clothes," added Candice.

"Can your bodies withstand being weightless for so long?"

Michael answered. "The International Space Station has taught us how to thrive in a zero G environment for indefinite durations. But as we said earlier, we won't really be weightless. The continuous thrust of the nuclear engine will provide acceleration to experience some force. Therefore, we will find living and working in Starship much easier than did the astronauts onboard the International Space Station."

"In addition, we each have adopted a regiment of daily exercise for an hour or longer," offered Candice.

"Can you shower up there?"

"Of course," said Candice. "We've learned a lot about personal hygiene on Space Station, such as taking sponge baths and recovering every drop of water for recycling. Sounds icky, I know. But our bodies will still remain attractive."

"Do you have enough space onboard for the rover, the food, the vegetable garden, the water, the fuel, the nuclear engine, and your living quarters? Won't it be cramped for such a long journey?"

"Actually, we have plenty of space," answered Kevin. The current Starship configuration is now 232 feet long and 30 feet wide. This gives us a volume of over 160,000 cubic feet which is much larger than the ISS. That would be equivalent to a 20,000-square-foot house with 8-foot ceilings if it were all on a single floor. Starship has many decks and compartments. Musk designed the Starship spacecraft to ferry one hundred people to Mars."

"Will your communication to Earth really be limited to short bursts since it will take four years to transmit the signal from the exoplanet to Earth and another four years to hear a reply?"

Jason answered. "You have the correct physics. Yes, communication with Earth will be severely limited, particularly as we leave the Solar System."

"So, how can we media follow your mission with such communication constraints?"

"You can't. NASA will provide you what little information they receive from us. They will also be tracking us, much as they have tracked the Voyager spacecrafts."

"What about radiation? How do you protect against it for such a long journey?"

"NASA has coated the inside of Starship with a new, highly protective shield," answered Michael. "It even works in the cupola. If the exoplanet does not have enough atmosphere or magnetic field to protect sufficiently from solar flares, we will build concrete-type habitats like NASA colonists constructed on the Moon."

"What if the exoplanet has a dangerous environment? Will you be able to return to Earth?"

"Our Starship has enough fuel to launch from the planet and return," answered Kevin. "And we think we will be able to use the water on the planet to make more fuel, just as we do on the lunar colony."

"Commander Owens, how reliable is the nuclear engine? If it malfunctions, can you return to Earth?"

"We do have the ability to return. It takes much less energy to do so from anywhere in the Solar System than to leave Earth. And it should serve us well if we have to leave the Proxima Centauri system to return. But it is highly unlikely that this will be necessary since it will have brought us all the way to fulfill our mission."

"We were told your team is cross training to enable redundancy to execute their duties. How can you learn so much in twelve months?"

"We all are very smart," quipped Jennifer.

"Is there anything you will miss on Earth?"

"Yes," answered Michael. "I will miss watching the Wisconsin Badgers and the Green Bay Packers football games."

"So, are you a cheese head?"

"Very much so. You can't be raised in Wisconsin without being a Packers' fan."

"I will miss playing golf," added Kevin. "But maybe when we land, I can reenact Alan Shepard's stunt of hitting a golf ball on the Moon. But we are not sure how far it would go in Proxima b's atmosphere and gravity."

"I may miss the excellent surgical facilities at Johns Hopkins, but NASA has provided such an elaborate collection of surgical and diagnostic paraphernalia that we may not want for anything," added Jennifer.

"I think I will have everything I need for artificial intelligence and robotics with the state-of-the-art computer and the amazing humanoid robot on board," said Jason.

"When you leave the Solar System, you will no longer have the Sun to generate power from solar panels. How will you generate power at that point?"

"The Sun's energy does not stop at the heliosphere boundary," said Jason. "In addition, Proxima Centauri's neighbor, Alpha Centauri A, radiates much more energy than our Sun."

"Starship also has special lightweight batteries that store excess power to be used during low solar and star radiation," added Kevin. "In addition, the effectiveness of the solar cells embedded in Starship's skin is amazing."

"Will your mission include observation of celestial bodies on your way out of the Solar System?"

"We certainly hope so," said both Michael and Jason. Kevin did not comment.

"Will you encounter any black holes?"

"No." answered Kevin. "The large black hole in center of the Milky Way galaxy is near the star Sagittarius-A, which is 25,000 light years away. However, in 2022 astronomers thought they discovered a small black hole just 5,150 light years from Earth with a mass only seven times that of the Sun, traveling away from us at 100,000 miles per hour. But this turned out to be not true. Another one, named GahaBH1, may be real. It is only 1,560 light years away and ten times the mass of our Sun. Some astronomers think there are one hundred million black holes in the Milky Way. But as far as we know none are close enough to matter."

"Eight years is such a long time. What if one of the crew members doesn't get along?"

"We throw him overboard," quipped Samantha.

"Actually, we have done everything possible to ensure crew compatibility," said Jennifer.

"Samantha is an accomplished guitarist and singer," added Kevin. "We will encourage her to perform a few concerts for our entertainment. But she is insisting that we all join in. She has downloaded a playlist that is phenomenal."

"I expect to convince each of my esteemed teammates to provide karaoke performances so we can all be entertained," added Sam with a big smile.

"We each have downloaded so many interesting books to our iPads, Kindles, laptops, and thumb drives," offered Jason. "It will take us years to read them all. I have personally downloaded the entire Wikipedia of two hundred gigabytes on my new thumb drive. It holds almost one terabyte."

"We will also continue our cross-training, learning each other's expert disciplines, which could occupy a great deal of our time," added Kevin.

"If your mission is successful and the exoplanet is hospitable, does NASA plan to send others from Earth?"

"You'll have to ask NASA," Kevin deflected.

"How can you build the infrastructure necessary to colonize?"

"We will have been trained to produce whatever we need upon landing using two 3D printers and additives onboard," answered Michael.

"NASA has demonstrated the efficacy of this technique both on the International Space Station and on the lunar colony," added Kevin. "We hope the regolith on the planet will provide sufficient properties to be used as basic material for manufacturing, just as it is on the Moon."

"How was this planet chosen?"

"I think we explained that," said Jason. "It is by far the closest exoplanet in the habitable zone."

"Could you abort and return to Earth from some point in the journey? Where is the point of no return?"

"Again, we answered that question," said Kevin impatiently. "We have sufficient fuel to return to Earth. And the nuclear engine can be used to kill off escape velocity at any point in the mission."

"Why are we doing this mission?"

"Space is the last frontier for us to explore. And exploration beyond the Solar System is the ultimate frontier. It is in our human makeup to explore. We hope to learn so much by exploring this planet and its own unique solar system."

"Will you six be awarded back pay if you ever return to Earth?"

"Hopefully, we won't have to find out. That should save the taxpayers money."

"Will your mission include search for and determining the source of Unidentified Flying Objects?"

"The Space Force now calls them Unidentified Aerial Phenomena and NASA has their own study of them. But so far, both have concluded that they are not of extraterrestrial origin. While it is not a mission objective to look for such flying machines, serendipitously perhaps we could observe one. But, based on these studies and conclusions, if we should encounter one, it will be highly unusual and unexpected."

"Have you paired up and chosen your mates yet?"

"We're working on it," smiled Jennifer. Kevin also grinned.

"What if a baby is born onboard?"

"We are taking precautions to prevent that from happening," said Dr. Thorson. "But we will manage it if it happens."

"Will your flight through the Solar System be like the Voyager's Grand Tour?"

"No, it will be quite different," answered Kevin." After launch, both Voyagers had no significant propulsion system onboard. They took advantage of the unique alignment of the four Gas Giants to use gravity assists to propel Voyager 2 to reach all four with no major propulsion, other than an attitude control system to properly align the communication antennae and reaction control jets for minor midcourse corrections. Subsequently, both left the Solar System for interstellar space. This particular alignment of the outer planets will not appear again until 2148. But we are planning to take advantage of the timing of the outer planets' current alignment to benefit from gravity assists in the vicinity of each planet. To do so, we will fly by each while they are traveling in the general direction of our intended trajectory so that the spacecraft velocity increase is en route to the next planet."

"So, we will also use the planets' gravity and momentum to speed up our Starship," added Jason. But the enormous difference is our onboard nuclear propulsion system. This will allow us to compensate for the slightly different alignment of the planets, as well as propel us through the Solar System much faster than Voyager. Also, our trajectory will differ from Voyager's. We do not have to rely on the curved Hohman transfer orbits dictated by the Sun's gravity, but the speed of our spacecraft enables our trajectory to be almost on a straight line compared to Voyager, thus shortening our distance and time between the planets."

"What if by the time your spacecraft approaches the next planet the gravity assists change your desired direction?"

"That is entirely possible," said Kevin. "We can use our chemical propulsion engines to correct for small deviations. We do not know precisely what acceleration we will receive from the nuclear engine. So, our flight time could differ from our original flight plan, causing us to arrive in the vicinity of the next solar system when the orientation of its planets must be considered. We would like to use these planets to do a reverse gravity assist to help slow us down. But they could take us in a slightly wrong direction. Knowing this, we would have to forgo the reverse gravity assist and continue to travel along the prescribed pre-determined path. Once we learn the position and velocity of these planets, our supercomputer will use this data to determine the necessary trajectory."

"Sounds complicated. How would this effect the mission?"

"We use these gravity assists to add velocity. Without these boosts, our slower speed would cause us to increase the time to get to the exoplanet. But not enough to compromise the mission. Conversely, using the reverse gravity assists as we fly by any exoplanet helps to slow us down as we approach our destination planet."

"Did NASA consider whether it is possible to go through a "worm hole' like Star Trek to reduce the flight time?"

Owens answered, "Now you are getting into science fiction again. While Albert Einstein and Stephen Hawking have indicated the possibility, even the probability of the existence of worm holes, using the

Eddington-Finkelstein metric to foreshorten space-time requires a connection between a black hole and a white hole to form a worm hole. At this time, we do not know whether white holes even exist. And as I explained earlier, we know of no black holes in our neighborhood."

"Any chance of your trajectory being effected by dark matter or dark energy?"

"NASA does not believe so. Scientists use these phenomena to explain the acceleration of the expanding Universe according to the Lamètrie principle and the Hubble constant, but the theoretical physicists have no proof that these are the correct explanations for the mystery of the galaxies' rapid expansion. So, I'm not going to worry about it."

"You must have a fear of the unknown."

Jennifer responded, "Sure. We don't know exactly what we may encounter. As they said in Star Trek, 'We are going where no man has gone before.' But that is part of the excitement. Wouldn't you all like to have that experience?"

Owens closed the press conference, promising a follow-up in the future before they launch. The reporters were abuzz with their plethora of information for their newspapers and nightly newscasts. They were not accustomed to getting such complete and intelligent answers to their questions. Their listeners were eager for news about these brave pioneers and their exploration of space. But they were disappointed with not being able to follow their journey in real time.

Afterward, Tony complimented the crew. "That was a great press conference. You all handled it well and know your stuff. One problem is the issue of crew diversity. It is causing a firestorm backlash. Both the media and the public criticize us as being narrow-minded and not accepting the tenets of diversity training. The press is hung up on this, so I'm recommending against another press conference. A follow up won't be necessary. Besides, you can't improve on this one."

Kevin retorted, "The press can go to hell. No more interviews by the media." He declined the offer from several magazines for personal interviews. Their attempts to entice went nowhere since money was irrelevant to them.

NASA continued to receive pressure from the media to select a more diversified crew. They ignored this push and issued the declaration that this crew is complete. But the criticism continued unabated.

"Be aware that this press conference likely made us famous," said Kevin to his crewmates. "So, watch out for the groupies who want to grab their fifteen intimate minutes with us. I heard that one hot beauty boasted that she slept with six of the seven original Mercury astronauts."

"And that is a bad thing?" smiled Candice. Jason frowned at her.

They soon found out Kevin was right. No longer were they able to drink and dine in private at their favorite hangouts. Young men and women crowded around them whenever they were in public, most just seeking their autographs. Others wanted more.

An enterprising local jeweler approached each crew member. She convinced each to be fitted for and to buy a 14-carat gold ring to enable a true marriage in space. She suggested they wear the ring on their right hand until they marry.

5. The Moon and Beyond

Kevin's extended launch reverie at Cape Canaveral launch pad 39A ends as Launch Control resumes the countdown. They are sitting in the Dragon capsule atop the 232-foot Starship spacecraft second stage, which in turn is mounted on the 230-foot Falcon 9 super heavy core booster. The launch stack is much taller than the mighty Saturn V that took astronauts to the Moon in the late 1960s and early 1970s. Elon Musk originally called it the "Interplanetary Transportation System." But NASA thought it needed a new name for this mission. Musk still preferred "Big Fucking Rocket." The upgraded booster sports 42 Raptor engines, each with a thrust of over ½ million pounds, while the Starship second stage now uses nine Raptors. At liftoff, the SpaceX superheavy booster delivers over twenty-one million pounds of thrust, more than 2.7 times the thrust of the powerful Saturn V.

Kevin's crew members are all wearing the latest technology spacesuits designed by two NASA contractors, originally for the Artemis Moon missions, based on 15 years of experimentation and research. But each proudly sports their mission insignia embroidered into the suit. The suit's modular design enables a more flexible fit for different body types, especially the women. They are lightweight, flexible, safer, and more comfortable than the previous generation designed by NASA in house. The new suits also replace NASA's extravehicular mobility units used for the previous two generations of EVA spacewalks. They evolved the design from lessons learned on Apollo, Skylab, and the International Space Station.

The launch window is only 180 minutes because they have to rendezvous with the lunar Gateway in its highly elliptical orbit to replenish the fuel they will consume during their launch and translunar burn.

While Kevin is excited, his crew members are nervous, some praying, as they await ignition. The 42 Raptor engines of SpaceX's booster ignite near the end of the launch window as almost one million people at the Cape watch intently. They have liftoff!

The entire world is viewing on television. The booster easily powers through maximum dynamic pressure as their G forces begin to recede, while their speed increases. Twenty minutes after liftoff, they experience main engine cutoff and a brief coast as the booster separates from the Starship second stage. The crew briefly floats against their constraints. The jettisoned booster begins its guided trajectory back for a powered landing to the launch pad. Then the Starship Raptor engines ignite, slamming the crew back into their couches. The shroud protecting the observation cupola retracts at 70 miles altitude as they leave the atmosphere. Starship completes the spacecraft's direct ascent to Earth orbit insertion at almost 18,000 miles per hour, shuts down its engines and begins its low Earth orbit two hundred miles high.

The crew is experiencing their first sensation of zero-G in orbit. They clamor up to the cupola for the spectacular view of the Earth, the clouds, the continents, and the oceans. They were anticipating the sight, but it is more fantastic than they imagined. Jennifer senses the idealized unity of the human race as she takes in the whole Earth. Michael is overwhelmed by the majesty of the sight and realizes the awe of what God has created. The entire crew experiences the perception of the beauty and fragility of the Earth.

After the first full orbit, Owens orders them back in their couches, buckling up with their restraints. He then commands the restarting of the nine Raptor engines. Once again, the crew is slammed back into their seats, experiencing more than three Gs of force. The engines cut off automatically at nearly 25,000 miles per hour beginning Starship's 66-hour translunar free-fall coast to Gateway. They experience zero-G floating again but are still held in place by their seat harnesses.

At this point, the crew happily unfasten their restraints, climb out of their bulky spacesuits and stow them in the cabin storage walls. They now subsist in the 50% oxygen, 50% nitrogen shirt-sleeve cabin atmosphere. They all scamper again into the cupola to watch the beautiful blue and white Earth get smaller. Most have mixed feelings at the sight. They observe the apparent unity of humanity on this small blue globe, yet know there is no unity, not across countries, not across America.

Kevin will not ignite the nuclear engine for any significant duration en route since he would only have to kill off the excess speed as they approach the Moon. The SpaceX Starship Raptor engines have enough fuel to later accelerate away from Gateway toward Mars before igniting the nuclear engine and more than enough to land on the exoplanet. But NASA and Kevin wanted a large fuel reserve in case they had not estimated enough for course corrections and landing on the exoplanet. After refueling, they will take on additional water at Gateway for the trip duration.

Once again, Kevin basks in the euphoria of weightlessness. The crew also enjoys it much more than the vomit comet. But each acknowledges the breathless view of Earth as they speed away. Although they were told what to expect of the majesty, it is much more than they anticipated, and it almost makes them nostalgic as they realize they will never see this beautiful planet again. They wonder whether their new planet will be anywhere near as impressive.

The nuclear engine was not yet ignited, not yet augmenting the cabin temperature with its warmth. Therefore, Starship rotates slowly on its axis to moderate the skin surface temperature between the warm sunshine and the cold shade enroute to the Moon. This also enables the solar panels to capture the Sun's abundant energy in its embedded fuel cells and to store the excess in its batteries.

The crew also loves the spacious accommodations in the Starship. The 180-foot tall, 30- foot diameter hull sports an inflatable table and chairs with a galley for their meals, separate sleeping pods together with space to store their clothes, a specially designed Peloton exercise bike with rotating screen and earbuds which they each use every day, a hydroponic garden which Samantha cultivates, a tiny music compartment with a playlist loaded into her new supercomputer laptop and thumb drive, a small guitar, a display monitor room with DVD player, and Velcro walls to house their toys when not in use. Stowed in fold-up position is also the two-passenger rover to be used on the exoplanet. It is positioned adjacent to the robotic humanoid.

Kevin tried out the nuclear engine for a brief time as they left Earth to verify its operation when they later left Gateway. It worked perfectly and gave them enough energy to reduce their travel time to the Moon by 11 minutes. Their human waste is processed for two essential functions: fertilizer for Samantha's garden and recycling urine into water. But they would not partake of her fruit and vegetables as yet. They were enjoying the thousands of ready-to-eat freeze-dried meals prepared by NASA.

Kevin relied on the capability of the powerful state-of-the-art flight computer for its navigation, guidance, and control software. He gave it verbal commands and it supplied verification of each command before igniting either the attitude control thrusters for small maneuvers or the Raptor engines for major velocity changes in the Starship trajectory. He could either enter the commands via a display keyboard or more often by verbal commands to "Orbiter," the name it recognized. The computer is triply redundant to ensure continuous operational performance and reliability.

Initially, Starship is in semi-continuous two-way communication contact with Mission Control at JSC relayed through its Goldstone Deep Space Network facilities around the world in Barstow, California, Madrid, Spain, and Canberra, Australia. This system also supplies spacecraft tracking of its position and velocity continuously once Starship is more than 19,000 miles from Earth. Communications and tracking are limited only by the distance from Earth since the signals travel at the speed of light. At lunar distance of 240,000 miles the signal is delayed by 1.3 seconds.

Candice showed off the zero-G yoga routine she had learned from the one demonstrated on the International Space Station. It included the ‘crescent moon’ and ‘star’ positions. She wanted the crew to follow her lead for the stretching and muscle toning benefits they provided, but most just stuck to the spacecraft’s Peloton exercise equipment.

“Starship, your trajectory continues to look good,” declared the Mission Control Capsule Communicator (Capcom).

“Roger,” replied Kevin. “We should be beginning to be affected by lunar gravity.”

As they approached the lunar sphere of influence 40,000 miles from the Moon, they were traveling at just over 2,500 miles per hour. Starship had lost 90% of its initial velocity achieved at engine cutoff because of Earth’s gravity. At this point Starship no longer lost speed from Earth’s pull but began to experience the gravitational attraction of the Moon, which caused their spacecraft speed to slowly increase. Starship must cross in front of the leading edge of the Moon to reduce some speed, in effect performing a braking maneuver. This is the inverse of a gravity assist.

“Houston. Approaching Gateway.”

“Roger, Starship,” responded Capcom, after the 2.6 second round trip communication delay. “Trajectory looks nominal.”

Kevin still had to restart the Starship Raptor engines to decelerate farther to rendezvous with Gateway, matching its orbital velocity at its pericynthion, 1,800 miles from the lunar surface. Its apocynthion was 4,200 miles away. He had learned how to rendezvous and dock with Gateway thanks to many hours in the training simulator. Still, he had difficulty docking because the orbital mechanics contradicted intuition. When Gateway was just a few hundred feet away and Starship was in the same orbit matching Gateway's speed, thrusting forward caused Starship's orbit to rise in altitude. When he decelerated Starship, the orbital altitude decreased, but the spacecraft increased in speed. He docked with Gateway after some difficulty with help from Orbiter. He then married the hose connectors to Gateway's fuel tank and water supply. The crew got a great look at the Moon from 1,800 miles.

"Wow, the Moon looks bigger than I expected," said Samantha. "Do they know how it was formed?"

"They learned from the rocks brought back on Apollo that it had to be formed less than 4.5 billion years ago, about 50 million years after the creation of the Solar System, when a giant Mars-sized body called Theia collided with the Earth during the formation of the Solar System," answered Jason. "The resulting Moon then receded to a wider orbit because of the tidal interaction with the Earth."

"I had no idea how many small craters pockmarked the lunar surface," remarked Candice. "I now have a better understanding on why Armstrong expended so much fuel to find a flat landing site in 1969."

"Apollo 11 landed with only 18 seconds of fuel remaining," added Michael. "I learned this in a book by a guy who was in Mission Control during the landing. The title is *My 36 Years in Space.* They were landing three miles downrange from their intended target on the Sea of Tranquility because NASA had modeled the Moon's gravity wrong. They thought it was a homogeneous spheroid like Earth. They didn't consider the mass concentrations produced by ancient meteor collisions when the Moon was still molten. The increased gravity of these mascons accelerated the lunar module faster downrange, resulting in the downrange error. This forced Armstrong to redesignate to a smoother landing site."

"Wow. I did not know that they came so close to disaster," said Jennifer.

"Well," added Jason, "that may not have been a disaster because they could have aborted, jettisoning the lunar module's lower descent stage, and igniting the upper ascent stage to rendezvous with the command module. But they also had computer alarms during descent, telling them that the computer was overloaded. Fortunately, the onboard computer software was designed to ignore the low priority programs, while executing the higher priority powered descent software, enabling a landing. Otherwise, they would have aborted, ignited the ascent engine, and rendezvoused with the command module."

"You guys sure know your Apollo history," complemented Kevin.

"I think I read every book on the history of Apollo ever written," added Michael.

"Kev, the Moon looks so desolate," said Jen. "What was it like to spend a month in the lunar colony?"

"I can't say it was enjoyable, having to be in my spacesuit any time I ventured outside the habitat. It was also extremely cold even though we had continuous sunshine because the Sun-angle at the South Pole was near horizontal, and the craters we visited never saw the Sun. But I learned so much about colonization. And the colonists were wonderful. I had a great mentor."

***

After fully replenishing Starship's fuel, oxidizer, and water tanks, Kevin used Starship's attitude control thrusters to slowly disengage from Gateway. He commanded Orbiter to restart three of its nine Raptor engines to accelerate out of lunar orbit and navigate to the fourth planet in the Solar System. Once again, the crew was pressed back into their couches. After the 50-second burn, Orbiter shut down the chemical engines. Starship sped away from the Moon at 21,200 miles per hour. Destination: Mars.

Kevin restarted the nuclear engine, benefiting from its slow, but sure, continuous acceleration. The crew reacted from a free-fall float to a gentle force against their Starship "floor." At first, Starship began to slow down because the gravity of the Earth and Sun at this position exceeded the acceleration caused by the nuclear engine.

"Orbiter, orient our velocity vector to perform a gravity assist from Mars at the closest approach of 100 miles altitude," he commanded. Orbiter's calculations carefully timed the restart to ensure the navigation to Mars met the necessary geometric and orbital constraints. After verifying Orbiter's response, he reported to Mission Control,

"Houston, we completed refueling, restarted the Starship's main engines for a 50-second burn, achieved lunar escape velocity, and restarted Starship's ion engine. We are on course for a Mars' gravity assist."

"Copy, Starship. Good luck and bon voyage."

The previous spacecraft probes, sent by several countries over the past decade to explore Mars, had no additional propulsion. Therefore, they took at least seven months to reach the planet. But the Starship nuclear engine reduced the transit time by almost two months, flattening the trajectory.

Enroute, the crew had plenty of time to explore each other. Kevin and Jennifer paired up as expected; Candice played with Jason; Sam and Mike tried each other on for size. All six thoroughly enjoyed their transit to Mars. No boredom here.

They also did other less strenuous activities. Each began to follow Candice's muscle toning and stretching regiment. Sam played her guitar, while the others tried to sing along. Jennifer and Jason began reading from their Kindles. Candice typed vigorously on her laptop, creating her journal. All spent at least one hour per day exercising with the Peloton. Jason always got excited watching Candice's nipples grow erect in her flimsy tee shirt as she peddled on the Peloton. Candice knew she had this effect on Jason and enjoyed teasing him. Their sleeping quarters became more active. NASA's stock of contraceptives was proving useful.

Kevin was happy to try out the mechanical razor designed by NASA. It gave him a smooth shave. Michael decided to use his as well. This made Samantha happy. But Jason kept his beard because Candice liked it. He trimmed it with scissors periodically. He looked quite debonair.

The crew eventually appreciated the fresh fruit and vegetables grown by Sam in her hydroponic garden. They enjoyed her tomatoes, lettuce, peppers, strawberries, and cucumbers. Not only were they healthy, but a welcome supplement to the packaged food provided by NASA. And they didn't have to add water to make them edible.

Starship reached the Mars' sphere of influence, 359,000 miles from the planet, traveling at a speed of almost 14,000 miles per hour. The Sun's gravity fighting against the constant, but low, acceleration produced by the nuclear engine caused the reduction in speed. Communication with Mission Control was minimal since it now took more than 15 minutes for the signal to reach Earth and another 15 minutes to reply.

"You continue to look good, Starship," barked the Capcom. "Any problems so far?"

"Roger, Houston," replied Kevin after the compulsory delay. "No problems, but it's weird trying to converse with a 15-minute delay. What's been happening on Earth since we left?" He attempted a bit of humor with a gross understatement: "The Internet is a bit faulty up here. I guess we miss it."

Thirty minutes later: "Okay, Commander. Here is a short news update. Mike's Badgers had a good season and will be playing in the Rosebowl next month. But his Packers are missing Aaron Rogers. The national debt continues to grow. Interest rates are rising, putting pressure on debt repayment. Let me know if you want political news. Everyone down here is interested in your odyssey and wishes you well."

"Thank you, Houston. Sure, lay all the news on us, including political."

Almost one hour later, later, he received the response. "Roger, Starship. Read you loud and clear. You are making substantial progress. Keep us abreast of your waypoints. Regarding your request for news, I'm afraid the politicians down here are at each other's throats even though the election is more than one year away. It gets uglier each year. And we call ourselves the UNITED States. Ironic, isn't it? But I suppose no worse than the United Nations."

Kevin did not respond.

By the time Starship reached the Mars' periapsis flyby point just one hundred miles from Mars, it was traveling at a speed of more than 20,000 miles per hour on a hyperbolic trajectory relative to the planet's coordinate system. Mars' gravitational pull had provided the acceleration, increasing Starship's speed. The spacecraft would have been on a perfect hyperbolic trajectory in free fall. But the speed produced by the ion engine slightly flattened this trajectory. Starship passed Mars on the trailing side in order to obtain the slingshot effect of a gravity assist to gain an additional 4,000 miles per hour. In effect, it gained some of the orbital momentum of the planet by flying by in the same direction as Mars' velocity.

From their 100-mile altitude, they received a wonderful view of the Mars' terrain from the cupola.

"Check out those mountains," said an excited Sam.

"Yes. Wikipedia says Mars' Olympus Mons is 69,000 feet, while Mount Everest on Earth is only 29,000 feet above sea level," said Jason. "And Mars' diameter is slightly more than half the diameter of Earth."

"And the valleys and deep gullies are so dramatic," exclaimed Candice.

"Scientists believe that Mars once had water," said Mike. "That looks like glaciers and dried-up riverbeds."

"Doesn't look like any water now," said Jen. "Those sand dunes, dust devil tracks, and dust storms must be every bit as bad as reported by NASA's Mars landers."

"Can you see the magnificent impact craters?" asked Kevin.

Candice was in awe. "Those photos we saw in training taken by the Mars Reconnaissance Orbiter don't do these views justice."

"Wish we could land and explore," added Jason.

"I'll pass," commented Samantha. "Those dust storms don't look like fun."

"I had hoped we could see Mars' tiny moons, Phobos and Deimos," said Jennifer. "I read that they are not spherical but resemble asteroids."

"Jen, if you look out the cupola to your left, I believe that speck is Phobos with the Sunshining on it," suggested Kevin.

"It is only 13 miles across and is less than 6,000 miles above the Martian surface," added Jason.

"Is that all it is?" questioned Jen. "Pretty disappointing compared to Earth's Moon."

"Houston, we just completed our Mars' gravity assist," said Kevin. "Received an additional kick of 4,000 miles per hour, just as planned. Next destination: the asteroid belt."

Less than seventy-five million miles from Mars, Starship entered the 175-million-mile-wide asteroid belt, shaped in a torus around the Sun between Mars and Jupiter. It is fifty million miles thick.

"Hey guys, we just entered the asteroid belt," announced Kevin.

"Are we in danger of a collision?" asked Candice.

"Not likely. The two Voyager spacecraft navigated through the 175-million-mile asteroid torus unscathed. Although there are thousands of measurable asteroids, the distance between major objects is so large as to not be considered a probable threat. The total mass of all the particles is less than the mass of Pluto, less than 4% the mass of the Moon."

"How large are these asteroids?" asked Samantha.

"From what I've read," answered Jason, "by far the two largest asteroids are Ceres with a 600-mile diameter and Vesta, 72% smaller."

"Orbiter's database contains the orbital track of the asteroids of substantial size," added Kevin. "So, Orbiter's navigation is able to avoid any sizable threats in our path."

But there were thousands of virtually undetectable dust particles in the asteroid belt. Traveling now at over 200,000 miles per hour, any size could present a threat.

Suddenly, Starship collided with one of these particles. It was not large enough to cause any real damage. It was too small for Orbiter to detect and sound a warning. But it did create an infinitesimal hole near the Dragon capsule.

"I think we hit one," suggested Samantha. She detected a minor leak in the Starship skin allowing their nitrogen-oxygen atmosphere to slowly escape.

"Not to worry," said Kevin. "We can fix it."

NASA had allowed for such a contingency by providing an adhesive patching material, much like duct tape with a mylar coating. It was not difficult for Mike to administer the patch.

"Patch complete. We are back in business," said Mike.

But the crew was apprehensive as they viewed two of these tiny particles zipping past the cupola. They were relieved to leave the asteroid belt weeks later without further damage.

"Glad to leave the asteroid belt in one piece," said Sam.

"Well, we could encounter some more in Jupiter's orbit," said Kevin. "They are called the Trojan asteroids and lie 60 degrees in front and 60 degrees behind Jupiter in its orbit. They lie at Jupiter's L4 and L5 Lagrange points with the Sun. They may be as numerous as the primary asteroid belt. Fortunately, Orbiter knows where any large enough to threaten us are. Some of the largest actually have moons of their own."

Jason said, "In 2021, NASA launched the Lucy Mission to explore these asteroids. They are believed to have been formed at the birth of our Solar System and so may tell us more about the ancient past. Lucy first made two flybys of Earth to obtain gravity assists, We are still learning from Lucy's findings. So far, we learned about their strange shapes. One is very oblate and looks like a hamburger. Another weird one looks like a chunk of it is missing from its southern side."

"Why did NASA call it Lucy?" asked Jen.

"I know!" exclaimed Sam. "Lucy is the name given to remains of the oldest humanoid discovered in 1974 in

Africa. The excavation team named her Lucy because they were listening to the Beatles' tune, *Lucy in the Sky with Diamonds* when they uncovered her. They think her bones are 3.2 million years old!"

"Pretty impressive, Sam," said Jason. "I didn't know you were also an anthropologist."

"No, I just am into the history of Beatles' songs."

"Houston," announced Kevin. "We are leaving the asteroid belt and bound for Jupiter. Keep up the information flow."

6. The Gas Giants

The nuclear engine kept accelerating Starship as it left Mars' periapsis and the asteroid belt. Kevin had to command Orbiter to fire the chemical engine for a short burst to align the trajectory toward their next gravitational assist target, Jupiter. Jupiter is almost five hundred million miles from the Sun.

Without the Mars' gravity assist and the nuclear engine, it would take six years to get there, as demonstrated by the two Voyager spacecraft, launched in 1977. But Starship reached Jupiter in less than nine months. Starship's speed exceeded 700,000 miles per hour as they approached the planet, easily overcoming the gravity of the Sun, thanks to the continuous acceleration.

"Starship," announced Mission Control. "Your flight path looks good. But again, the news is not so good. The political infighting is worse than last year. Both parties are blaming each other for the degrading economy, the high inflation, the growing national debt, and the high interest rates. The stock market is suffering. The public is not optimistic. We expect voter turnout to be low. So sorry for this report, but you requested not to sugar-coat the news."

The crew reacted with mixed feelings: sad about the US economy; happy they were not there to live through it.

"We are approaching one of Jupiter's trailing Trojan asteroids," announced Orbiter. "Even though its orbital plane is inclined 20 Degrees to Jupiter's plane, our trajectory is almost crossing it. So, we should get a good look although our closest approach will be more than 1,000 miles."

"This was one of Lucy's targets," said Kevin. "The bigger one is called Patroclus: the smaller companion four hundred miles away is Menoetius. They are pretty large for asteroids: seventy miles and sixty-five miles in diameter, respectfully."

"They look pretty strange from here," said Jen. "They are not round. Do we know how they were formed?"

"We can only guess. Maybe from a collision. Lucy may be able to tell us."

Michael asked Kevin, "Will we visit the extraordinary natural satellites in the rest of the Solar System?"

"I'm afraid not. Our mission is to colonize the exoplanet, not explore and tour along the way. In 2016, NASA's Juno mission found that Jupiter has at least seventy-nine satellites, so we may see some of them en route or on our way out."

"But these four major Galilean satellites are fantastic and have never been seen by people up close," said Jason. "We can add so much to our knowledge of the Solar System."

"I saw them with the University of Wisconsin telescope in my astronomy class as an undergraduate," added Mike. "You can see them with a pair of binoculars from Earth if you can hold them steady enough. Kevin, let's at least get close enough to see one or two."

"That would cause us to expend precious fuel and delay our trip by days. I am sorry, but I have to decline."

"Orbiter, how much fuel would we need to fly by some of Jupiter's and Saturn's exotic moons?" asked Candice.

"It could reduce our reserve supply by several percent," replied the supercomputer almost instantly.

"I thought that was why we refueled at Gateway, enabling us to manage contingencies along the way," added Mike.

"The answer is no," declared Kevin. We will use the gas giant planets of Jupiter, Saturn, Uranus, and Neptune to increase our speed via gravity assists as we learned in our training and the mission plan. If we can see some of their satellites along our trajectory, it will be serendipitous. My decision is final."

This was not well received by the entire crew. Jason and Mike were steaming but decided not to challenge Kevin further … at least not now.

"Starship, you are making excellent progress and time," said the Mission Control Capcoms. "Keep up the excellent work.

"Now for more news. The good news is that the midterm election is over and both parties are starting to play nice, at least in the House and the Senate. The 2030 election resulted in victories for the Democratic Party. So, the Republicans are screaming that the millions of illegal immigrants have voted in the states allowing mail-in ballots. They point to the precincts in which more votes were counted than registered citizens. Audits have not overturned any results as yet."

"Crewmates, should we respond to this news?" asked Kevin.

"Another downer," remarked Jen.

"But what can WE do?" asked Sam.

"Right. There is nothing we can say or do that will affect anything," said Michael.

"Don't bother responding," agreed Candice.

"I'm glad I am here and not down there," repeated Jason.

Kevin proceeded to enable Orbiter to navigate on the planned trajectory to Jupiter's gravity assist peri-apsis.

After Mike cooled down, he confronted Kevin. "Let's talk about this decision about our odyssey. We will never again have an opportunity to explore these fascinating features of the Solar System. We have plenty of fuel for contingencies, and we don't mind a delay of a few days to have experiences of a lifetime."

"I agree with Mike," said Jason.

"Kev, let's talk about it," added Jen.

"No, I am the commander and am not changing my mind. This is important."

"Yes, you are the commander, but we are all equals here," chimed in Candice. "This is not the military. And you are not a king, Kevin. We all have an equal stake in this mission."

"I'll tell you what, I will contact NASA and get confirmation that my decision is correct.

But it will take more than 90 minutes to get a response from Mission Control from here.

"Houston, some of our crew want to fly by some unusual satellites of the giant outer planets. I told them this could jeopardize our primary mission by depleting some of our fuel reserve and adding unnecessary delay. Please explain the folly of this idea."

Eighty-five minutes later the Capcom responded. "Commander Owens, this is your decision. While there may be important scientific information to obtain from such a flyby of these exotic satellites and would provide a rare opportunity, your primary mission is to colonize the exoplanet. If the fuel and time would permit such excursions, you might consider it only if it would not threaten the primary mission."

"Such excursions could compromise our mission, so we will not do it," declared Kevin.

"Wait a minute," replied Jason. "We have plenty of fuel in reserve and delaying our landing by a few days does not compromise the mission."

Candice repeated her argument that this is not a military crew and Kevin is not a king. She continued, "Therefore, we are all equal and this is a democracy. Let's vote on it. How many want to explore the Solar System without expending fuel that would compromise the mission?"

Jason, Mike, Candice, and Jen all raised their hands. Sam was not sure, so did not vote.

Kevin replied, "Tell you what. I will acquiesce to keep unity in our crew, but I will use Orbiter to plan the trajectory to minimize the fuel expenditure required to fly by a few satellites, but not many."

Orbiter's database and algorithms were programmed with the best navigation solutions for obtaining multiple gravity assists of the gas giant planets, based on a PhD thesis submitted to the University of Glasgow in 2010, *Global Optimization of Multiple Gravity Assist Trajectories* by Matteo Ciriotti.

"Orbiter, what is the fuel requirement to visit any of the four Galilean satellites: Callisto, Ganymede, Europa, and Io, while preserving the Jupiter gravity assist?"

After a short pause, Orbiter responded, "Our current trajectory will allow us to fly by Callisto, the outermost moon, and to the innermost, Io, with little propellent expenditure. But the other two require major changes in our gravity assist trajectory."

"Okay," said Kevin, "based on Orbiter's calculations, we will fly by Callisto as we approach Jupiter and fly past Io on our way out."

Mike and Jason were overjoyed with this plan and wanted to hug him. But Jason was disappointed not to see Europa. The NASA spacecraft, Juno, had discovered interesting ridges and grooves 93 miles wide and 150 miles long across Europa's frozen surface, which might have been caused by material bubbling up from the depths of Europa's ocean. There was scientific speculation that this was a possible place for primitive extraterrestrial life. NASA launched another probe in 2027, planning to arrive in 2033, to observe it.

Kevin ordered Orbiter to perform the mid-course correction to pass within fifty miles of Calliso's trailing edge as it reached the left side of its orbit around Jupiter. Since they were still over one hundred million miles from Jupiter, Orbiter only had to fire the main engines for three seconds to achieve the flyby.

Callisto is the farthest of the four Galilean satellites 1,130,000 miles from Jupiter. It is the third largest satellite in the Solar System and is the size of Mercury. It is almost 40% larger than Earth's Moon. It is tidally locked to Jupiter, so, like Earth's Moon, the same hemisphere always faces the planet. By flying by on the trailing edge, Starship gets an additional gravity assist, but it is small.

As Starship approached Callisto, the crew observed a stark contrast from the surface of Mars.

"I don't see any mountains or valleys," remarked Sam.

"Right," commented Jason. "This moon's surface consists of impact craters; thousands of them. I read that it is the most heavily cratered body in the Solar System. None of the craters are caused by ancient volcanos."

"Callisto looks like it has some kind of atmosphere," said Candice.

"Right again," lectured Jason. "This atmosphere of carbon dioxide and oxygen is extremely thin, allowing us to see the surface. But it does have a strong ionosphere caused by solar radiation."

"It is believed that Callisto has a major subsurface ocean, perhaps 60 miles under the surface," continued Jason.

"Half the surface is water ice and half rock," added Mike. "I think those bright spots you see are forms of ice."

Sam observed, "The leading hemisphere appears to be darker than the trailing half."

"Correct again," lectured Jason. "They think the leading edge has more sulfur dioxide, while the trailing hemisphere has more carbon dioxide."

"You guys DO know your stuff about these satellites," remarked Kevin, beaming with pride.

Satisfied and smiling, the crew relaxed to complete their inbound trajectory to Jupiter's periapsis, taking advantage of the Oberth effect, wherein use of a reaction engine at higher speeds generates a greater change in energy than its use at lower speeds.

"Kevin, we will experience Jupiter's periapsis 2,000 miles above its gaseous surface to avoid its atmosphere," announced Orbiter.

Jason again educated the crew. "Jupiter's distance from the Sun is 484 million miles. It takes almost twelve years to orbit the Sun. But the Jovian day is only ten hours long."

The pull of the Sun's gravity at this distance was greatly reduced. Starship reached a speed of 700,000 miles per hour as the Jovian gravity pulled them toward Jupiter's periapsis.

Candice noticed a faint, reddish ring around the planet as they approached and pointed it out to Jason.

Jason responded, "Actually, we are seeing the two outer 'gossamer rings' of Jupiter's four ring circus, first discovered by Voyager 1 in 1979. We will see a bright, thin 'main ring' next, followed by the inner bluish 'halo ring.' The rings are mainly composed of dust particles. The gossamer rings start at 135,000 miles from Jupiter; the main ring starts 77,000 miles out; the halo ring begins at 73,000 miles. The rings are only about thirty feet thick."

Candice was impressed by Jason's knowledge.

As they approached closer, the entire crew was astonished by the size of Jupiter's Great Red Spot. Of course, they knew about this perpetual storm on the planet, causing the red appearance, but the enormity of it was mind blowing. It was larger than the whole of Earth, with winds blowing more than two hundred miles per hour.

Once again Jason began his astronomy lecture. "What we are seeing is only Jupiter's upper atmosphere. It is composed of 90% hydrogen and 10% helium with traces of other gases, such as ammonia. Although its total atmosphere's mass is 71% hydrogen and 24% helium, and 5% many other elements, we can only see the 30-mile layer of ammonia, which contains the Great Red Spot. All its visible storms occur in the thin upper atmosphere layer."

Mike added, "Scientists believe Jupiter has a solid core the size of Earth, based on its enormous mass. Yet, it produces a gravity only 2.4 times that of Earth. But its mass is more than twice the mass of all the rest of the planets combined. Its diameter is eleven times that of Earth."

Approaching closer, they remarked on the colorful stripes and bands in the visible atmosphere. Jason explained, "The light-colored bands indicate the gas rising upwards; the dark bands indicate the gas sinking downwards. The colors are due to their chemical composition and temperature differences."

"I am impressed," said Kevin. "When did you learn all this?"

"Partly from my astronomy classes, partly because I had been studying this stuff since I was accepted to the mission crew. Wikipedia is a treasure of facts."

Starship was traveling at over 720,00 miles per hour as it approached Jupiter's periapsis. The gravity assist added another 16,800 miles per hour.

Near their closest approach the crew marveled at the beautiful green and red rings of aurora borealis around both the North and South Poles. "These are much more intense than the northern lights I saw in Alaska," remarked Sam.

Kevin explained, "The Sun's charged particles interacting with the strong magnetic field around the Poles produce the auroras. We may see them at Saturn, Uranus, and Neptune as well."

"What fun, being the first humans to witness these beautiful effects in our Solar System," Candice responded. "More great stuff for my journal."

As Starship began its modified outbound hyperbolic trajectory, Jason continued his lecture. "Io is slightly larger than our Moon, with a diameter of 22,000 miles. It is in a circular orbit 250,000 miles from Jupiter, the innermost of the Galilean satellites. It has the highest density of any satellite in the Solar System with almost no water. It whips around Jupiter in only 42 hours."

"There it is!" shouted Jen. "And I think I see a volcano!"

"I'm sure you do," responded Jason. "It has more than four hundred of them. They come from tidal heating within its interior. The plumes are sulfur and sulfur dioxide and can reach three hundred miles above the surface."

"Look at the mountains!" said Jen.

Jason commented, "There are over 100, some as high as Mt. Everest."

"Are those lava flows?"

"Yes. The many colors are sulfur allotropes. The flows can be hundreds of miles long. They cause a very thin atmosphere."

"Too bad we can't see Ganymede," said Jason. "It is the largest moon in the Solar System, even bigger than Mercury. Astronomers think it has an underground ocean with more salt water than the Earth's surface water. It actually has its own magnetic field, which lies within Jupiter's magnetic field. We could see the spectacular northern lights in its thin atmosphere. It has both dark and bright terrains. Scientists think it may have some form of life.

"Another weird moon is Europa. It orbits Jupiter twice for every single orbit by Ganymede. It also has a giant subsurface water ocean one hundred miles deep with cryogysers that erupt on the surface and are visible. It too could support life. Europa has a relatively smooth surface. If it is successful, early next year NASA's Europa Clipper spacecraft will learn a lot more about this fascinating moon. We would know a lot more today if NASA's funding cuts didn't cause more than a three-year delay in its development and launch. It received a gravitational assist from Mars in 2028 and another from Earth in 2029. NASA plans for it to orbit Jupiter in an elliptical orbit and make forty-four close flybys of Europa. We'll have to learn about the results from Mission Control next year."

Mike added, "You know the Galilean satellites were a big part of the adoption of the Copernican model of the Solar System in the 16th century. Prior to that, most believed Earth was at the center with Sun, Moon, and planets orbiting Earth."

***

On the way out, Orbiter announced it had a surprise for the crew. "We will be passing within twenty miles of one of the many irregularly shaped outer moons. It is moving in a highly elliptical orbit well out of the plane of Jupiter's orbit. Its orbit is retrograde to the orbits of the Galilean satellites. You should be able to view it from the cupola in two minutes."

"Wow. Thank you, Orbiter," said Candice as she climbed into the cupola. "This moon is ugly; not at all spherical."

Orbiter elaborated, "It is called Thyone, part of the Ananke group of retrograde orbits. It is only two miles wide and has an orbital period of six hundred days. All these satellites are in my database, enabling me to carefully avoid any collision."

"And I thought YOUR skill was navigating us through these dangers, Kev," challenged Jen.

"Well, Orbiter is giving me a little help," responded Kevin with a big smile.

The nuclear engine was now accelerating Starship at two miles per hour for every second that it emitted its nuclear ions, causing a force of 0.1 G on their bodies. The Sun's gravity was impeding their acceleration less with increasing distance from the Sun.

"Houston, we are leaving Jupiter and bound for Saturn," announced Kevin.

On to exciting Saturn. After the amazing adventure of the Jupiter flyby, they were anticipating even more from this exotic planet. But patience was required, even traveling at almost one million miles per hour. The Cassini probe took seven years to get to Saturn; Starship expected to make it in another two months, thanks to the continuous acceleration from the nuclear engine.

The crew was still bubbling about their encounter with Jupiter and its moons.

"I didn't realize this trip would be so adventurous," offered Candice. "I thought the eight-year travel time might be repetitive, tedious, and even boring."

"You'll get your chance for boredom when we leave the Solar System and reach interstellar space," remarked Kevin.

"Just seeing Jupiter up close was a sight to last a lifetime!" echoed Sam.

"I can't wait to see what Saturn offers," added Jen.

"It could be orgasmic," chimed in Kevin, with a big smile. "But we have approximately two months to digest the images before we reach Saturn."

"Orbiter, which of Saturn's moons are we able to fly by with minimal propulsion from our chemical engines to alter our planned trajectory to a Saturn gravity assist?"

"Commander, we can see Enceladus on the way in and Iapetus on the way out with less than ten seconds of main engine burn to redirect our primary trajectory," responded Orbiter. "But if you want to see Titan, it will require a burn of more than 90 seconds."

Mike interjected, "Saturn has eighty-three moons at last count, but most are small and not very interesting. Titan is the second largest natural satellite in the Solar System, only exceeded by Ganymede. And it has a thick atmosphere of nitrogen with an atmospheric pressure

50% higher than Earth's. So, we probably couldn't see the surface which is primarily water ice. But I think Enceladus and Iapetus will be more spectacular."

"Let's go for it, Orbiter, but forget about Titan," commanded Kevin.

Once again, the crew grew excited with anticipation.

"Starship, congratulations on your successful encounter with Jupiter," interrupted Mission Control. "Once again, your trajectory is nominal.

"On the home front, no new news as the country is resting after the midterm election."

"Houston, we are ready to proceed to Saturn. We will swing by Enceladus on the way toward periapsis and visit Iapetus as we exit."

Ninety-seven minutes later, Kevin received a reply. "Roger, Commander Owens. Keep us informed on your progress."

"Continue to keep us abreast of the news back there," requested Kevin. "I know the delay makes it difficult to communicate, but you're our only source." He decided to make another smart remark: "We are a little too high to catch the Internet transmission," he quipped. "Nor does the news media quite make it out here. I never thought this would happen, but I actually miss the Internet and the news."

After reorienting the attitude control system, Orbiter executed an 8.6 second burn of Starship's Raptor engines.

Four hours later, Mission Control reported, "Okay, Starship. Here is your news update as requested, but it's not all good. President Putin is rattling his saber again; this time threatening to annex the Baltics and return them to the Soviet sphere. The US and NATO warned that they will defend Latvia, Estonia, and Lithuania as promised in the NATO alliance, but we are all concerned not to back Putin into a corner. He again threatened to use tactical nuclear weapons.

"Also, the demonstrations in the largest American cities are becoming violent again, challenging law enforcement.

"At NASA, we are pleased that the Artemis colonization of the Moon is proceeding well. The Chinese are continuing to make progress with their space program. They expect to land a crew on the Moon in a few weeks."

Kevin responded, "Thanks for the update, Houston. Continue to tell us what is going on and give it to us straight. We can handle it.

"I'll report again when we get to Saturn."

Although the next two months were uneventful, the crew continued to buzz about their Jupiter encounter and anticipate the wonders of Saturn. They were not disappointed. It was an incredible sight. So beautiful, unique, and majestic. Their planned trajectory took them through the seven main concentric rings of Saturn. Actually, they did not really see seven distinct rings. The widest ring had a diameter three times the diameter of Saturn. They flew through the Cassini Division between the outer A ring and the much wider inner B ring. Since these rings lie in the plane around the equator and since Saturn's axis of rotation tilts 26.7 degrees to its orbital plane, the crew had a magnificent view of the rings as they entered and exited the Cassini Division. The 9,000-mile-wide A ring started 82,000 miles from the gas giant's gaseous surface. There, the 3,000-mile-wide Cassini Division began, ending at the start of the B ring. The B ring was 15,000 miles wide. Adjacent was the C ring which extended to just 45,000 miles above Saturn's gaseous surface. None of these main rings was more than ½ mile thick, the thinnest only sixty feet. The crew was thrilled to fly between the rings.

"What are these rings made of?" asked Candice? "They seem so fragile."

"They are almost 100% ice crystals," answered Jason. "They actually have their own atmosphere of oxygen and hydrogen, sparse as it is. The Sun's radiation ionizing the ice particles creates this. Most of the particles are less than an inch wide, but a few are as large as three feet."

"Are we in danger of colliding with them?" asked Sam.

"Not really," said Kevin. "They have such a low density that their mass is irrelevant to Starship, even at our high speed."

"How were the rings formed?" asked Jen.

"The scientists and astronomers are unsure," said Jason. "There are several conflicting theories."

"How about the gaps like the Cassini Division?" asked Jennifer.

"Some were cleared out by Saturn's many moons. They think the moon Mimas, which circles the planet once per Earth day, cleared out the Cassini Division. It is tidally locked with the planet. Mimas has a diameter of 240 miles."

"How many moons does Saturn have?" asked Sam.

"As I said earlier, the latest count is eighty-three. Of these, twenty-nine haven't even been named yet. But only thirteen are larger than 30 miles. Oddly, several of them orbit retrograde; that is opposite to the rotation of Saturn itself."

"How is that possible?" asked Sam.

"It is believed they are captured minor planets or pieces of them."

"So where is Enceladus?"

"Orbiter, where is Enceladus?" asked Kevin.

"We will encounter it in nine minutes," responded Orbiter.

"There it is!" shouted Mike.

"Impressive," said Sam. "Are those volcanoes that I see?"

Jason began another discourse. "Not exactly. They are geysers of water ice. We think Enceladus has a liquid water ocean under the surface. In addition, it is mostly covered by fresh, clean ice. This is why it is so reflective of sunlight. The moon is either heated by the decay of radioactive elements in its core, or by tidal flexing caused by Saturn's gravity. It is speculated that some form of life is supported, habitable life. It does have an atmosphere composed mostly of water vapor."

"I wish we could land and explore to determine whether it does support life," said Sam. "Is there any way we can check it out?"

"Not without landing," answered Kevin.

"It looks immense. How big is it?" questioned Jen.

"Enceladus' diameter is 310 miles. But it is only Saturn's sixth largest moon. Its orbit is 143,000 miles from Saturn," Jason responded.

"We're approaching Saturn's periapsis and will pick up our gravity assist of 12,600 miles per hour," announced Kevin.

"Saturn's diameter is 72,000 miles," reported Orbiter. "We will be 1,800 miles from its gaseous surface to stay above the dense atmosphere and its winds, which can reach 1,000 miles per hour."

"There is an interesting, but little-known fact about Saturn," said Jason. "Its density is 30% less than water."

"So, if we could throw it into a cosmic-sized ocean, the planet would pop up like a cork?" asked Jennifer.

"Well, maybe not as dynamic as a cork. But, yes, it would definitely float."

"Is Saturn the only planet whose density is less than water?" asked Candice.

"Yes, at least the only one in the Solar System. Just another thing that makes Saturn unique."

Much later, after the gravity assist flyby, "We are now approaching Iapetus," announced Orbiter.

"Look!" yelled Jen. "Iapetus is not spherical. It's not even an ellipsoid. It looks like it has a big bulging waistline and squashed poles. It looks like a huge walnut."

"Well, it is the third largest of Saturn's moons," said Jason. "It has a diameter of nine hundred miles. It is 80% ice, based on its low density. That fat waistline is due to an eight-mile-high ridge with twelve-mile-high peaks around its equator. When we get a little closer, we may be able to see very large impact craters. Several are over two hundred miles wide. In fact, the largest is Turgis, 360 miles wide. This guy orbits Saturn at over two-million miles altitude, more than twice as far from the planet as Titan, and it takes almost 80 days to complete the orbit. Because of its distance and 15-degree inclination to Saturn's equator, we can see the rings more clearly from here. It is tidally locked by Saturn and therefore one side always faces the planet."

"Wow. You're right," interjected Candice. "The view of the rings is spectacular; much better than when we flew through them."

"Look at the color difference. One side is a dark reddish brown; the other is really bright," observed Jen. "It looks like the two sections of a tennis ball."

"Yes. Scientists think the dark material is residue from the evaporation of ice from the warmer surface areas. The gigantic orbit and its synchronous motion with Saturn's rotation results in prolonged exposure to the Sun, heating the surface temperature, causing the temperature difference between the Sunlit side and the colder dark side."

"Okay, gang," said Kevin. "Say goodbye to Saturn. We're on our way to Uranus."

"Houston, we are now leaving Saturn. The sights were spectacular," reported Kevin.

More than two hours later Starship received Houston's response. "Thank you, Starship. You are looking good. Any problems, Kevin?"

"Also, here is your news update as you requested. We certainly have problems here on Earth. Putin has formed a stronger alliance with China. He continues to threaten to invade the Baltics. NATO and the United States verbally say they will continue to support them, but they are concerned about Putin starting World War III if we follow through with using our aircraft to defend against his aggression. Putin has threated to use his nuclear stockpile which exceeds ours. At this point we don't know whether XI Jinping will enter the conflict. This does not look good."

Kevin responded, "That is horrible news. Our prayers are with you all. It is so frustrating to deal with these time delays. But we here are fortunate not to be exposed to the conflict in real time. But isn't there anyone there with the cojones to stand up to Putin?

"Meanwhile, to respond to your question, our problems are small and dealt with easily. We are enjoying the ride and the observations of our Solar System."

Kevin turned to his crewmates, "I can't believe that our leaders are permitting Putin's threats to make us appear weak and enabling our enemies to intimidate us."

"But they ARE weak and feckless," responded Mike. "They proved that when we abandoned Afghanistan and surrendered it to the Taliban, allowing terrorists to once again have a safe haven to coalesce and threaten us."

"Right," said Jason. "I can't believe what is happening to our country and the world."

"I'm happy to not be present as our weakness and capitulation are exposed to the world," said Jen.

"I agree," echoed Sam. "Do we really want to continue to hear such news?"

"Watch out for Xi Jinping," warned Candice. "He is observing this weakness and will be encouraged to go after Taiwan."

Almost three hours later, "Roger, Starship. Thank you for your prayers. We continue to debate our options regarding the international situation.

"Continued good luck to you and your crew on your journey to swing past the ice giants for your final gravity assists."

7. The Ice Giants

"Orbiter, are there any interesting moons around Uranus that we could swing by without significant use of our Raptor engines?" asked Kevin.

"Uranus has twenty-seven moons; five are large enough to have interesting features, but we can only fly by one without a large burn. That is Miranda, the smallest of the five."

"Okay, Orbiter, take us to Uranus' gravity assist with a swing by Miranda," commanded Kevin.

By now they were traveling over thirty million miles per hour and were able to reach Uranus's sphere of influence in less than two months.

Along the way they never tired of listening to Sam's guitar talent and her beautiful voice. Again, she encouraged all to join her. By now they knew most of her songs.

At 60,000 miles from Uranus, they encountered the first of the Planet's 13 rings. Although dark and faint, it was a few miles wide, enabling observation by the crew. The rings were comprised primarily of ice particles 5 to 50 feet in diameter.

On the way toward the periapsis of Uranus, Orbiter took them within two hundred miles of Miranda. They only had a glimpse of the satellite due to their high rate of speed. Nevertheless, the crew were amazed at the major geysers shooting out from the moon.

"It seems to be coming out of Uranus' orbital plane," observed Candice.

"Yes," said Jason. "That's because Uranus' axis is tilted 97 degrees from its orbit plane. Miranda is orbiting Uranus' equator, as are all five of the major moons."

"The surface seems to be broken," added Jennifer.

"On the way here, I read about Miranda on Wikipedia," offered Jason.

"You can't get Wikipedia from here," commented Michael. "You need an internet."

"Right. But, as I told you last year, I downloaded the Wikipedia description of the entire Solar System into my iPad's flash drive before we left Earth. It is a wealth of information.

"This is an interesting, unusual, and mysterious moon, even though it is only 280 miles wide. Gerald Kuiper discovered it in 1948. In 1986, Voyager 2 flew within 18.000 miles of Miranda, which provides much of our knowledge of this moon. It orbits in tidal lock of Uranus 77,000 miles from the planet with a period of only 34 hours. Its orbital inclination is more than four degrees, ten times that of Uranus' other major moons."

"Candy and Jen, you both made good observations," continued Jason. "The expansion of subterranean ice expanding and splitting the surface ice caused the uneven terrain. Miranda has huge canyons hundreds of miles long. It has the largest cliff in the Solar System -- 12 miles high. Miranda's low density indicates it is at least 60% water ice."

"Wow, that was a quick look at Miranda," said Candice. "I wish we could slow down to get a better look."

"We are not far from Uranus' periapsis where we will pick up another 9,000 miles per hour relative to the Sun with another gravity assist," announced Kevin. "It doesn't sound like much, but it's free and will reduce our flight time in interstellar space."

"Hey look!" exclaimed Candice. "Uranus has a bunch of rings like Saturn. I see a blue one and a red one as we approach."

Jason again displayed his amazing knowledge. "Those are the two outer rings. Uranus actually has thirteen rings caused by its magnetosphere. The rest are not colorful.

"Uranus' atmosphere is the coldest in the Solar System, as low as 371 degrees below zero. It is in sparse layers of water and methane and appears aquamarine in color, which is what we are seeing now. It has windspeeds of up to 560 miles per hour at different latitudes. They think this may be what causes the dark spot that we are seeing now. Check out the bright white solar cap on the South Pole compared to the rest of the atmosphere."

"So, Uranus has a giant spot like Jupiter?" asked Candice.

"Yes, but not as big as Jupiter's. It is a vortex measuring 1,100 miles by 1,900 miles, caused by its churning atmospheric winds. Its atmosphere is very similar to Jupiter, but colder. Its layered cloud system consists of methane at the top and water vapor at the lowest layer.

"As we saw earlier, its axis of rotation is tilted sideways, almost lying in its orbital plane. It takes 84 years to complete a single orbit around the Sun because it is almost two billion miles away. Because of its tilted rotational axis, each pole receives 42 years of sunlight. Its rotational rate is 17 hours, but at some latitudes the rotational rate of the atmosphere is only 14 hours due to the fierce winds. This difference in the brightness and color of the cap around the South Pole is probably due to the strange axis orientation."

"Why is its axis so screwed up?" asked Jen.

"No one really knows," answered Jason. "It is possible this happened billions of years ago during the formation of the Solar System when a large protoplanet collided with it."

"Does Neptune have this crazy axis tilt?"

"No, Uranus is unique."

"Houston, we're leaving Uranus and Miranda and are proceeding to Neptune," reported Kevin.

More than five hours later, Kevin heard Mission Control respond, "Thank you, Commander. Good to hear your voice. You are on schedule with a good trajectory. God speed, Starship.

"On the home front, nothing much has changed regarding the Russian aggression, but neither has anything been resolved.

"Domestically I am sorry to report that we are experiencing another wave of crime in the large cities. The local politicians have not been successful in hiring and training enough qualified police to solve the problems. Also, our education system has degraded, based on test scores. We are currently ranked 35th in the world in math and 27th in reading. The public education system is accused of emphasizing Critical Race Theory and social issues over the basic fundamentals. Meanwhile, China ranked near the top in both critical skills, but educators dispute this due to China's unorthodox method of scoring."

"Houston, feel free to include any good news along with this depressing stuff," transmitted Kevin.

"Okay, Orbiter," said Kevin. "On to our last gravity assist at Neptune. Anything interesting along the way?"

"Yes," responded Orbiter. "Neptune's largest moon, Triton, is almost on our flight path."

"Great. Go for it."

"You mean fly by it, correct?"

"Yes, of course."

Almost two months later, Orbiter announced, "We are approaching Triton."

"This is a rare opportunity," offered Jason. "Triton is in a retrograde orbit, traveling clockwise around Neptune. Its inclination to Neptune varies from 127 degrees to 173 degrees. According to Wikipedia, it likely had to be a dwarf planet captured from the Kuiper belt. It is similar to Pluto, but much larger, not formed from Neptune's dust. It is geologically active with five-mile-high geysers of nitrogen gas and has a thin transparent atmosphere."

"Yes, I can see it!" exclaimed Samantha. "It looks pink."

"It is by far Neptune's largest moon at 1,700 miles in diameter," continued Jason. "The other thirteen moons are tiny and irregularly shaped. It is tidally locked to Neptune with a rotational rate equal to its orbital period. It revolves around Neptune in just 141 hours."

"How did Wikipedia learn all this about such a distant planet?" asked Sam.

"We can thank Voyager 2's 1989 Grand Tour flyby for most of this."

"It looks brighter than other moons we've encountered and appears sort of reddish color," said Sam.

"Yes, it is. Its icy surface reflects at least five times the sunlight that Earth's Moon does.

"There may be a form of primitive habitable life on Triton."

"Once again, I wish we could land there and find out," said Samantha.

"Well, that might be fun to explore, but its surface temperature is almost absolute zero. I'm not sure our spacesuits could keep out the cold very long."

"We are moving so fast that I can barely see some surface ridges and valleys," continued Sam.

"You have a good eye, Sam," said Jason. "It has few craters, which you can't see anyway at this speed, indicating it is not very old. But you should be able to see the caldera caused by one of the largest volcanos in the Solar System. Check out its 1,200-mile-long volcanic dome. Try to see the two giant cryolava lakes."

"This is all good stuff for my journal," remarked Candice, joyfully.

"And if you could transmit it to Earth somehow, you may be adding to our knowledge of the Solar System," responded Jason.

"Triton is no longer in our field of view, but we're approaching Neptune itself," announced Kevin.

"Hey, it looks a lot like Uranus," said Jennifer.

"Well," responded Jason, the Wikipedian, "it is almost the same size and color, but it is a darker blue, and its axis is right-side up. Its axial tilt is 28 degrees, much like Earth's. Also, you can see the weather patterns in the atmosphere. Check out the cool bands of methane clouds. See its giant dark spot? It is like Jupiter's Great Red Spot and like Uranus's dark spot. The winds are up to 11,000 miles per hour, the strongest in the Solar System. Its outer atmosphere is 360 degrees below zero."

"So, we might freeze our tushes off if we visit," quipped Jen.

"That would be a big loss for me," smiled Kevin, again admiring Jen's firm derriere.

"You probably can't see it, but it does have a faint ring system," continued Jason.

"Neptune is almost three billion miles from the Sun and takes 165 years to complete one orbit. So, its four seasons last 40 years each. Yet, its day is only 16 hours long."

"Okay, Orbiter, take us to Proxima Centauri," commanded Kevin.

Orbiter realigned the spacecraft's Raptor chemical engines and began a three second burn to achieve the required trajectory.

"Houston, we are leaving Neptune's sphere of influence and entering the Kuiper belt," announced Kevin. "We completed the burn and are on our way out of the Solar System. Our crew is healthy and in good spirits."

Almost nine hours later, he received the Capcom reply. "Roger, Starship. That's good news. Your state vector and trajectory look good. Be careful through the Kuiper belt. Wish we could communicate in real time.

"More national news. The House has a bill that would allow every resident to vote, citizen or not. If this passes, the administration's open border policy will pay dividends. The two-party system may be dead in the United States.

"America's debt is now approaching thirty-four trillion dollars. With the interest rates now up to 8%, our credit rating has fallen to BBB, according to Moody's. There is concern that the Treasury may default on our bonds. The interest rate is hurting the economy. Our

Gross Domestic Product continues to fall. The stock market is tanking. Tax increases will be necessary to make our bond debt payments. There continues to be major unrest in our cities.

"I am sorry to report all this unwelcome news. Are you sure you want to hear this stuff?"

"Wow," said Sam. "I don't need to hear any more of that."

"No, we need to hear it," responded Candice.

"That is so depressing," said Jason. "But we need to know."

"Why?" chimed in Michael. "There is nothing we can do."

"Still, we need to keep the communication link open," declared Kevin.

"I agree," said Jennifer. "It's hard to hear. But we can't put our heads in the sand."

Kevin responded, "Keep the news flowing and don't hold anything back. But add any major golf tournament news."

"Kevin, I'd like to hear about the Packers' and Badgers' football," requested Mike.

"Houston, also give us all the bad news about the fortunes of the Green Bay Packers' and the Wisconsin Badgers' football," kidded Kevin.

"Anything else, guys?"

"Yes," said Jen. "Any medical breakthroughs?"

"Houston, add any significant progress or breakthroughs in the medical field, please."

***

"Team, we are now entering the Kuiper belt and will fly through it for the next two billion miles," announced Kevin. "This will be different from flying through the asteroid belt because of the great distance between the objects. The Kuiper belt is thirteen times wider. Orbiter knows the location of the 35,000 Kuiper Belt objects larger than 60 miles wide, so we will not be in danger of hitting them, even though this is hundreds of times the number in the asteroid belt."

"Orbiter, will we be able to see Pluto?" asked Michael.

"Yes, and you may first be able to see another dwarf planet, Eris, which is almost as large as Pluto and is actually more massive," answered the supercomputer. "Other known dwarf planets include Orcus, Haumea, Quaoar, and Makemake."

"Kevin, I learned of a Caltech discovery named 2003 UB313, a Kuiper belt object likely 50% larger than Pluto," proclaimed Jason. "They think it is three times as far from the Sun as Pluto, but its orbit is inclined 45 degrees to the main plane of the Solar System. Can we fly near it and observe it?"

"No. According to Orbiter, it would require too much fuel for the plane change."

"Commander," interrupted Mission Control. "regarding your request for sports news: Tiger Woods notched his 85th PGA victory in the US Open at age 53. He is now only one major victory behind Jack Nicklaus. The Green Bay Packers under their new quarterback finally beat the San Francisco 49ers to get to the Superbowl. But they lost a close one to the Buffalo Bills, 27-24. The Wisconsin Badgers advanced to the National Finals but lost to the Baylor Bears on the last play of the game."

"Regarding national news. No progress in the cities or the congress."

Many hours later, Michael observed, "That must be Eris."

"Wikipedia says it is 1,400 miles in diameter," said Jason. "It orbits the Sun in a giant eccentric ellipse. Its aphelion is over nine billion miles from the Sun; its perihelion is 3.6 billion miles. It takes over five hundred years to orbit the Sun, but its rotational rate is only 26 hours. It even has its own moon, Dysnomia."

"I don't see a moon," observed Mike.

"That's because it is only 400 miles wide and we are traveling too fast."

"We are now approaching Pluto," announced Orbiter.

"Look," said Michael. "It has huge mountains."

"I think that is the Tanzing Montes range," declared Jason. "They reach up to 20,000 feet above this small dwarf planet. We probably can't see it now, but Wright Mons is an ice volcano three miles high and ninety miles wide. It may still be active, probably formed from a volcano ejecting subsurface water, then frozen as it hit the extremely cold surface."

"Wow! There's another one," exclaimed Mike.

"That's probably Piccard Mons," added Jason. "It's even bigger; four miles high and 150 miles wide. These are unique to Pluto. Although they looked snow-capped like Earth's mountains, Pluto has no atmosphere with water vapor. So, its 'snow' is frozen methane, rather than ice crystals. But their mountains are composed of rock-hard water ice. Astronomers think these mountains can only be less than one billion years old, while Pluto was formed some 4.5 billion years ago."

"How did those bodies get way out here?" asked Jen.

"Experts really don't know. Astronomers hope to learn more from wide field survey telescopes now being built. Apparently, there was not enough matter out here to form a real planet. Astronomers believe the total mass of the Kuiper belt to be approximately one percent that of Earth."

"Jason, when you are through, please let me borrow your iPad," said Mike. "I would like to learn more about the Solar System."

"Me too echoed," Candice. "What an adventure we had through the Solar System. We should have videoed it. But it gave me great material for my journal."

"Well, we are still in the Solar System and will be for a couple billion miles," said Kevin.

"Spectacular," said Jennifer. "What a thrill! No one has ever seen this before," she reminded herself – and everyone else.

"And it was all created by God," added Michael.

"Are you sure?" asked Candice.

"Yes, I am. Years ago, I was agnostic. As a man of science, I easily dismissed Genesis. But after college I wanted to be sure. I read the entire Bible. There had to be a good reason that billions of people accepted it. I also read books by scholars and journalists on the subject. In addition to the Bible, two books made an impression on me: *The Genesis Question* by Hugh Ross and *The Case for Christ* by Lee Strobel. They changed my life. I then believed. I accepted Christ as my savior and joined a Presbyterian church. I now know that as a believer, I will have everlasting life after I die."

"Michael, do you believe everything in the Bible?" asked Samantha. "Do you believe it is the word of God?"

"Oh, I still do not believe everything in the Bible. Although God inspired the writing, there are several passages in the Old Testament I believe are in error. They were written by many people in ancient times who were fallible and knew much less than we do today. "

"Which passages are flawed?"

"Genesis for one. God's creation of Earth. I believe God created the Universe with the Big Bang. But Genesis's timeline is flawed. The Universe was created fourteen billion years ago. But I do believe that God did create the Earth, and after billions of years, He created humans in his image 200,000 years ago, not 6,000 years. Also, I have a problem with the angry, jealous, vengeful God of the Old Testament who destroyed much of humanity with the great flood. And why did He order genocide for the Canaanites, rather than just winning them over? I don't believe God killed the man for touching the Arc of the Covenant, trying to keep it from falling off the cart in 2 Samuel. And why did He prohibit Moses from entering the Promised Land after his great leadership of the Israelites?

"But I do believe the New Testament description of the gentle, kind, forgiving God is correct and refutes these flawed passages of the Old Testament.

"I believe in the Apostles Creed: that Jesus is the son of God, that He was sent to Earth to save us from our sins; that He died to accomplish this and arose from the dead."

"I would like to read your copy of the Bible," requested Sam.

"Let me read it when you finish," echoed Candice.

"I think Michael is spot-on right down the line," offered Jason. No one disagreed.

"Please lend me your copy of *The Case for Christ*," requested Jennifer.

"Well, they are both in my Kindle, so you'll have to take turns," responded Michael.

"Your faith is strong, Mike," continued Samantha. "You told us about it last year, but please elaborate. I want to know more."

"My belief in God has developed only the past few years. But it gets stronger every day.

"When I completed my formal education, I began seeking the Truth. I read the Bible, but being a man of science, I was initially turned off by the description of the creation in Genesis as I told you about. As I read further and became immersed in the stories and their lessons, I accepted most of the written Word, especially the New Testament. Then I read those books I named earlier that convinced me of the Truth:

-That God did create the Universe and life within it,

-That Jesus Christ is the Son of God,

-That He walked the Earth 2,000 years ago, died on the cross to forgive our sins, and rose from the dead."

"But, from what you discussed earlier, you don't believe that the Bible is the Word of God. You said you didn't believe all the passages," said Jennifer.

"Right. Obviously, the beginning of the Solar System, creation of the Earth, and the creation of life in Genesis are chronologically inaccurate. "

"So, you believe in creationism, rather than evolution?"

"No; I believe in both. I believe God originally created the heavens and Earth, but evolution took us from there to where we are today."

"Do you believe in the Big Bang theory?"

Yes, I certainly do. I believe that God created the Big Bang. No scientist today even claims to know exactly how the Universe was created fourteen billion years ago. And no one has a clue what happened during the first 380,000 years after the Big Bang. God is the only answer. All the matter in the Universe was instantly created from nothing except energy."

"Think about it. Before the Big Bang, the Universe did not exist. It was created out of nothing that humankind can postulate. Then it expanded during the first 380,000 years until the galaxies were born. Then the Universe expansion accelerated, rather than slowing down. Our knowledge of physics says that not only should gravity cause the expansion to slow down, but eventually to contract, resulting in a "Big Crunch." Scientists had to invent dark energy and dark matter encompassing the Universe bubble to explain this perpetual accelerated expansion. This beautiful, amazing Universe had to be created by God."

"What do you think, Jen?"

"I have been searching for the truth. I have read most of the Bible but still have reservations. I think the book you recommended, the *Case for Christ,* may help."

"I think it will. I am no missionary or evangelist, but I have become a strong believer, in part due to this book."

Samantha's physical and emotional attraction to Michael grew stronger with this exchange. She also reinforced his interest in her.

Candice and Jason listened to the dialog between Michael and Samantha with great interest. They too looked forward to reading *The Case for Christ*.

***

Months later Kevin announced, "We are now out of the Kuiper belt, We are approaching the edge of the Solar System, and entering interstellar space. We may actually make it to our destination in eight years after all. We reached interstellar space in only 20 months. From this point on, we will make better time because the Sun's gravity will have a negligible effect. Our current speed is fifty million miles-per-hour."

Jason added, "So, we are nearing the heliosheath, some eleven billion miles from the Sun. This amounts to approximately 123 AU, where an AU is the distance from the Earth to the Sun. At this point, the interstellar wind balances the solar wind. We should be feeling the turbulence of a bow shock wave as we encounter some interstellar dust. We should celebrate leaving the Solar System for interstellar space."

"Who brought the champagne?" asked Candice.

"I knew I should have brought a couple of bottles," said Jen.

8. Interstellar Space

Four months later, Kevin alerted the crew, "We are about to enter the Oort cloud. The time lag on a signal from here is almost six months. Jason, what can you tell us about the Oort cloud?"

"Not much. It differs from the Solar System which lies in a flat disk around the Sun. But they think the Oort cloud is an enormous thick sphere surrounding the Sun. We know very little about it. We think there is an inner cloud, a flat disk-shaped region starting some two hundred billion miles from the Sun. We are near that point now. But its spherical outer cloud extends as much as twenty trillion miles further. Therefore, it is well outside the Solar System. We think the objects in the cloud may be no larger than ten miles wide. But we think there are at least one hundred billion of them, and maybe as many as trillions; we really don't know. In general, they do not truly orbit the Sun because the Sun's gravity is almost negligible at this distance. Yet, they believe the cloud to be the source of most of the comets. Although the objects may be small, we can speculate their total mass could be as large as the mass of Jupiter because we surmise that there are so many. There is so little we know because Voyager will not get here for another three hundred years. So, we are the first to be on a journey "where no man has gone before" to quote Star Trek."

"That gives you an idea of our tremendous speed," said Kevin, "currently 150 million miles-per-hour, more than 20% the speed of light."

Jason continued. "Astronomers speculate that these objects may be 'leftovers' from the formation of the Solar System 4.6 billion years ago. Not even the James Web Space Telescope can see these tiny objects."

"Right," added Kevin. "And Orbiter does not have them in its database. Therefore, we have to rely on our radar to spot any threat in time for Orbiter to exercise its collision avoidance system to keep us safe. Fortunately, the Oort density is so sparse that we don't expect to experience close encounters. By the time we exit the Oort cloud we will be more than halfway to our exoplanet."

Suddenly, the attitude control jet blasted to avoid a collision with a rocky object just a few miles from them.

"Nice job, Orbiter," congratulated Kevin. "Nice to know the collision avoidance system works."

"Hey, is that a comet?" asked Sam, excitedly.

"I think it is," responded Jason. "It is clearly traveling away from the Sun on a nearly parabolic orbit.

Its two tails come from the outgassing of ionized molecules and dust. The solar wind produced them."

"But its tails look like they are in front of the comet, not behind its motion."

"Yes. That is because the solar winds always push the tails away from the Sun independent of its direction of travel. We are very lucky to see it. Comets are quite rare. And most travel in highly eccentric orbits that take many centuries before they return."

Kevin announced, "NASA is having trouble tracking us at these distances. I will begin using our compact infrared laser directed at Earth. It will transmit a short nanosecond burst of megawatt energy whenever I fire it. This will be possible for NASA to detect and learn our position. It will augment our tracking via the Deep Space Network."

***

Samantha was taking her exercise turn for one hour on the Peloton when her mind began to wander. She was thinking back to her days as a high school cheerleader. She was a popular girl, whose beautiful face and figure attracted many boys, some of whom she dated. But something was missing in her physical attraction to them. She enjoyed the petting and let them take it to third base, but she never went all the way. Something held her back. Then, one evening after practice, one of the girls came on to her. Instead of being repulsed, she participated. Afterward, she realized she must be bisexual. But she didn't want to be a lesbian. She wanted to have a loving husband and have children -- many of them. She turned away from homosexuality and went further with her boyfriends, enjoying the journey. But she was concerned of her lesbian liaison and hid it from her psyche. She certainly was successful in hiding it during the crew interview and hid it from Michael very well. She was very much in love with this wonderful man and knew she would be happy with him for the rest of her life.

***

Everyone looked forward to their biweekly two-hour poker games. Six was an excellent number of players. They played dealer's choice wherein each person became dealer in turn, and he/she would declare the type of poker game they would play for that hand.

Kevin's favorite was Texas Hold-em. After the small ante by the player to his left and a bigger ante by the next player, each player received their two cards from the dealer. Then each of the rest of the players would either match the bigger ante or fold. Optionally each player could raise the ante at his turn. Again, each player in turn would have to match the raised bet or fold. Then, the dealer would bury a card face down followed by three community cards (used by each player) face up, called "The Flop." Then a round of betting commenced. The first player to the left of the dealer had the option of checking (no bet) or betting. If he/she bet, the succeeding players had to match the bet, raise it, or fold. If he/she checked, the next player would have the option of checking or betting. Each also had the option of raising, but the maximum number of raises was three. Then the dealer dealt another card face down followed by one card face up. Another round of betting ensued. Finally, the dealer dealt a final card face down and one face up. The player's hand always consisted of all face up cards in addition to his two original cards, but a maximum of five of the seven would count. The final round of betting ensued. If the player did not fold his/her hand, he/she would show their two cards. He/she could use any of the seven cards to create his/her best poker hand. The winner held the best five cards of the seven. The best possible hand is a *straight flush*: five consecutive cards of the same suit. The highest, AKQJT, is called a *royal flush* and cannot be beaten. Next is four of a kind; again; the denomination governs, ace being high. Next is a *full house* consisting of three of the same denomination plus a pair of two others of the same denomination. Next is a *straight* of five cards in sequence of any suit. Next is *two pairs*; then *one pair*. Finally, if no one has any of the above configurations, the hand with the highest card wins, ace being highest.

Omaha was Jennifer's favorite because there were usually two winners: a high hand and a low hand. As a result, fewer players folded. Again, it started with the person left of the dealer submitting the low ante and the adjacent player submitting the high ante. The dealer dealt four cards to each player, followed by the first-round betting identical to Texas Hold-em. Then, like Texas Hold-em, the dealer faced three community cars (The Flop) and a round of betting ensued. The fourth and fifth rounds were also like Texas Hold-em. Each player MUST use two cards from his/her hand and three of the five from the board to assemble the poker hand. There were two winners, high and low, unless the low five cards did not contain all five denominations lower than a nine. In that case the high hand won the entire pot. If another hand matches the same winning result, the pot is split equally.

Michael's favorite poker game was Seven Card Stud. Each player was dealt two cards down and one up. A round of betting followed. Each of the next three rounds was dealt face up, followed by a round of betting. The final round was dealt face down, then betting. Any players who have not folded used their best five cards to win.

Jason liked a variation of Seven Card Stud called Seven Card Stud High-Low. There are almost always two winners because there was no restriction on determining the best low hand, unlike Omaha.

Candice liked wildcard games. Her favorite was Seven Card Stud – Deuces Wild. This differed from Michael's favorite in that any deuce can be declared to be whatever card the player wishes. This often resulted in much higher hands than would exist without the wild card deuces. If a hand with its wild cards resulted in five of a kind (same denomination), it won over a *straight flush.*

Samantha's favorite was a crazy wild card game: Seven Card Stud – Low Hole Card Wild. This differed from Michel's and Candice's favorites in that the lowest hidden (hole) card in each player's hand was wild as well as any others of the same denomination whether showing or not. Often, this required an extremely high hand to win and could be exciting. The seventh card to each player is dealt face down; therefore, it could become the lowest concealed hole card.

Occasionally, just for variety, Kevin would choose a new game called 2-2-1. Like Omaha, it began with each player receiving four concealed hand-held cards, but the common cards were all dealt face down: four on the top row, two on the bottom -- Like Seven Card Stud – High Low, there were always two winners. After the first round of betting, the dealer would face one of the cards, followed by a round of betting. Each of the subsequent rounds, facing another card, enabled another round of betting (or checking), but the last round faced the last common card: always in the lower row. Each player was required to use exactly two cards from his/her hand of four, two cards from the top row of four, and one card from the bottom row. This game always produced the largest pots since there were two guaranteed winners and seven rounds of betting. Until the last round one could not be sure of the value of his/her hand, notwithstanding the hand of his/her competitors.

Since money was of no value apart from Earth, they simply used poker chips; then kept a running total of the results each session.

The best player was clearly Jennifer. "What is your secret?" asked Kevin.

As if on cue in a musical, Jen began singing the chorus to the Kenny Rogers' song, *The Gambler*. "You've got to know when to hold-em; know when to fold-em."

Sam harmonized, "Know when to walk away; know when to run."

They all joined in, "You never count your money when you're sittin' at the table. There'll be time enough for countin' when the dealings done."

Smiles all around.

Candice thought playing for chips was a little dull. She suggested strip poker. Jason was okay with that, but the others nixed the idea. Then she suggested the option of the winner of the session demanding that the loser do whatever he/she desired.

***

One day Kevin complained, "These muscle cramps in my legs I get almost every night are driving me crazy. They are so painful. But they almost disappear when I stand up and put weight on my feet."

"That's a common complaint of either too much exercise or not stretching enough," said Candice.

"I don't think I'm over exercising, and I do stretching exercises daily before I work out."

"It could also be caused by dehydration or a lack of magnesium in your system," said Jennifer.

"I'll give you some magnesium pills, but just take a small dose as they may have side effects."

Two days later, Kevin announced, "The magnesium is working – no leg cramps. Thanks, Jen."

A few days later, Kevin said, "Well, no leg cramps, but the side effects are worse. It is screwing up my digestive track. I'm going to stop taking the pills. I prefer the pain."

Candice offered, "I can fix the leg cramps. Jason had them for a while until I worked on his muscles. Come to my pod tonight before you go to bed, and I'll show you."

Kevin appeared in her pod that evening. Candice was wearing her famous tee shirt and shorts.

"Okay Kevin. Take off your clothes, cover your butt with this towel, and lie on my bed on your stomach. I'll show you what a great physical therapist I am."

Kevin did as she commanded, while Candice lubricated her hands with oil.

She began massaging the soles of his feet.

"Mmmm. That feels wonderful. I didn't realize my feet were so sensitive."

"The soles of your feet are the most sensitive part of your body," she responded.

Candice began kneading his calves.

"Your fingers are so strong."

"Attributes of my profession.

"You have major knots here. This should eliminate your cramping."

He was loving it. But as she worked her way up his thighs, he was getting aroused. Part of him wanted her to stop, but he was enjoying it too much. He suppressed a moan.

"Okay, Kev. Turn over so I can work on the tops of your legs."

He turned carefully, keeping the towel covering his pelvis. But his arousal was now obvious.

"Uh, oh. Wow! It seems I created another problem for you, Kev. But I can fix that too."

Before he could react, Candice quickly removed the towel and began "fixing."

Again, part of him wanted her to stop. Instead, he found himself begging her **not** to stop. There was no turning back. She was so skillful.

But she also was becoming excited. Her natural juices began to flow.

Many erotic minutes later, Kevin was satiated. But now a wave of guilt overcame him.

Kevin reached for his tidy whities and began dressing. "Candy, we can't do this again."

"Why not? Didn't you enjoy it?"

"More than you know. And thank you. But I don't want to mess up my relationship with Jen."

"Wait a minute, Kev. It is now your turn to tend to my problem. A little reciprocity, please. I know you are capable."

"Aren't you concerned about your relationship with Jason?"

"Well, yes. We have a great sex life. But we are not married. I'm not sure we want to be monogamous for such a long time. A little variety can spice up this odyssey."

"Let's discuss this tomorrow. I know I will sleep well tonight, thanks to you."

This brought a big smile to Candice's face. But she was still frustrated and made a beeline to Jason's pod.

The next day, Kevin requested a meeting of the crew. "I'd like to discuss our views on sexual monogamy. All six of us are mature adults who are free to choose our sexual activities. However, in forming a colony, I believe we should be raising our families on a solid foundation. To me that means a monogamous relationship with our chosen mate. One of the major problems with our communities on Earth is the absence of children raised by both their father and mother.

Children raised by a single parent tend to be unsuccessful in their relationships, their self-esteem, and their careers. Too many end up in a life of crime or worse. We need to establish the proper monogamous foundation for our community here on Starship. If not, our relationships could become complicated and confusing. What do you think?"

Mike immediately responded, "I couldn't agree more. Sam and I have discussed this thoroughly. We are committed to each other for life. In fact, Kev we would like you to marry us."

"Is that what you want, Sam?"

"Oh, yes! Can you do it, Kev?"

"As the captain of the ship I have the authority, and I would love to perform the ceremony."

"Kevin and I have developed a great relationship and are sexually exclusive," said Jen. "I wouldn't have it any other way."

Candy thought, *Well, I guess my choices are severely limited.* She decided to stay silent on any contrary opinion. Actually, she really agreed with Kevin's logic.

Jason offered, "What you say makes a lot of sense to me. Candy and I will discuss it."

A few days later, Candice and Jason approached Kevin. "Kev, we've discussed monogamy and concluded we want to be devoted to each other for the rest of our lives. Please marry us. We've talked to Mike and Sam and would like to have a double wedding."

"Great! I will perform the ceremony whenever you like."

"We don't want to wait. How's tomorrow?"

***

Each exchanged beautiful vows of love, fidelity, faith, devotion, and respect. Obedience had long since disappeared from old traditions. It was replaced with equality, much to the gratification of Candice. Each of the four removed their gold wedding band (that they bought before they left JSC) from the ring finger of their right hand and presented it to Jennifer. She, in turn, gave them to their new spouses, who placed the rings on their betrothed's ring finger, concluding the formal ritual. The four newlyweds walked down the 25-foot isle to the traditional wedding processional songs played on Sam's laptop of *Greensleeves, Because,* and *Endless Love.* Then they all danced to the rest of Sam's playlist: *Just the Way You Are, You Are My Reason, I Get to Love You, A Thousand Years, How Long Will I Love You,* and *Can't Help Falling in Love.*

They celebrated with a "wedding cake" salad which Samantha made from red romaine lettuce, tomatoes, radishes, Swiss chard, Chinese cabbage, chili peppers, cucumbers, and peas. She left out the onions and garlic for obvious reasons.

After the ceremony, Jennifer cornered Kevin. "Well, Kev, what about us?"

"I love you, Jen. There is nothing I'd rather do than spend the rest of my life with you. Will you marry me?"

"Of course. But who can perform the ceremony?"

"As my deputy, Mike has the authority. I'm sure he would do it."

A few days later, Michael performed the ceremony for Kevin and Jennifer. Thus, this now clarified the future of the colony.

They delivered some appropriate vows. "Jennifer, you are my light. And it doesn't matter what happens in this life as long as you are with me. I, Kevin Owens, take you, Jennifer Thorson, just as you are -- loving everything that I know of you and excited about what I have yet to discover. I vow to respect and trust you as long as we both shall live through everything that life brings our way."

"And I, Jennifer Thorson, take you, Kevin Owens, to be my best friend and my captain … to be your anchor and your sail, your starboard and your port as long as we both shall live."

Mike, Candice and Jason joined Samantha in two recessional songs as the wedded couple skipped back down the short aisle: *The Best Day of My Life* and *All You Need is Love.*

The crew entertained themselves with a ritual of removing the partitions between their sleeping pods. They were three happy couples. They learned they could be creative with their sexual romps in the 0.1 gravity environment. Monogamous sex became their number one activity with great entertainment value in interstellar space. But they still were careful to use the contraceptives supplied by NASA.

The second-best activity remained the singalongs with Sam and her guitar. They were becoming proficient singers. She taught Michael to join her in harmonizing with his deep bass voice and her soprano, while the others carried the tune with their baritone and contralto voices.

Each has their favorite tunes: Sam's are country-western, particularly those by Blake Shelton, Johnny Cash, Patsy Cline, and Miranda Lambert. Jason's favorite vocalists are Josh Groban, Ella Fitzgerald, and Andrea Bocelli. Kevin likes the classic pop ballads by Whitney Houston, Michael Bublé and Johnny Mathis. Candice prefers punk stars like Adam Lambert, Grace Slick, and Elton John. Jen prefers the stylings of Barbra Streisand (although not a fan of her politics), Aretha Franklin, and Taylor Swift. Mike is into Amy Grant, Celine Dion, and Elvis Presley, whom he enjoys imitating. -- Quite an eclectic genre collection. All loved the music from the old Broadway plays, especially the classics, like *Oklahoma, The Sound of Music, South Pacific, West Side Story, My Fair Lady, Les Misérables,* and *The Phantom of the Opera.* They all learned to love singing. They usually closed their homegrown concerts with their own rendition of Frank Sinatra's *Fly Me to the Moon.* Their fond reminiscences of Sam's and Mike's karaoke duet in the Seabrook bar more than two years ago was their bonding memory.

▪▪

Starship was partly into the Oort cloud traveling at 22% the speed of light when a collision warning blasted through the ship. Suddenly the attitude control jets instigated a precipitous lurch to the side, knocking the crew off their feet. Candice's arm slammed into the wall of the ship. Then they heard a thud. A different warning sounded, together with flashing lights.

"What was that?" exclaimed Jason. "I think we had a collision!"

"We are losing our air!" yelled Michael.

"No, any effected compartment will seal itself off from the rest of the ship for now to prevent widescale loss of pressure," reasoned Kevin. "But get into your space suits until we can verify the problem. I need someone to do an EVA with me outside the Starship to assess the damage."

"I'll do it, Commander," volunteered Michael. "I was pretty good at it in the simulator."

Kevin was proud of the lack of panic displayed by the crew members. But he was concerned about Candice's arm as Dr. Thorson began to examine it.

"Crew," announced Kevin. "I'm going to shut down the nuclear engine while we do the EVA to examine the damage. So be prepared to be in freefall again like we were on our way to the Moon."

Kevin shut down the nuclear engine to eliminate the 0.1 G acceleration, returning Starship to a freefall coasting state. He gave Michael a hand-held light beacon. He grabbed the laser tracer contraption from the Deck 9 spacecraft tool hold.

Suited, Kevin and Michael slowly entered the egress tunnel and fastened the Starship tethers to their suits. Kevin sealed the egress valve and began to loosen the external hatch. The air quickly rushed out as the hatch began to open. Kevin and Michael used their EVA training to extract themselves carefully and deliberately from the spacecraft, experiencing the absolute void and blackness of space. Michael shined the beacon along their way as they used the Starship handrails to explore the spacecraft's skin until they located the breach. Both experienced the difficulty of inching their way along the handrails. They used their spacesuit communication intercom to converse.

"Here it is. This is the compartment where we store our collapsed rover. We must have collided with a rocky Oort object," said Kevin. "It created a gaping hole of nearly three inches by two feet. Fortunately, it didn't hit any of our attitude control thrusters. We have to tear the dangling metal off the hole to repair it. The tiles of the Thermal Protection System are gone. But the inner blanket is still there."

"We can repair this, and I can fix any damage to the rover," said Michael, puffing with exhaustion.

Michael and Kevin struggled for 40 minutes to carefully remove the metal without piercing their gloves.

Kevin used the laser tracer to precisely map out the path of the hole, creating a computer-aided design template. Tired, they returned to the airlock, resealed the hatch, and re-entered the tunnel. Kevin brought the metal skin piece into the main cabin and located one of their sophisticated rapid prototyping 3D printers. Jennifer was surprised and pleased when she learned that this was the same RP printer used in hospitals to create replacement body parts and artificial tissue.

Michael fed the laser tracer CAD results into the printer with the acrylonitrile butadiene styrene thermoplastic, blended with the skin composite material to generate a molten stainless-steel patch with only four passes of the printer head. Three minutes later it hardened to produce the patch that perfectly matched the hole. They returned to the tunnel and released the hatch. They carefully climbed out and inched their way back to the damaged breach. Michael fitted the patch into the hole and applied the special type of duct tape liberally, sealing the intersections.

"Looks good," complimented Kevin. "Nice job, Mike."

Kevin and Mike carefully returned to the airlock and breathlessly rejoined the crew. Kevin restarted the nuclear engine, restoring their artificial gravity.

"Did you smell that strange odor of space after we climbed out of our spacesuits?" asked Kevin.

"Yes. It is still on our suits, helmet, and gloves," said Mike.

"It is a pleasant, metallic smell that only emits from space. Very weird. I don't understand it, but almost every astronaut that has done an EVA has experienced it."

Jen asked, "So, Kevin, what happened? Are we in trouble?"

"We apparently collided with a small rocky object in the Oort cloud. But the damaged compartment automatically sealed itself off from our Starship to prevent our cabin air from escaping. Mike patched the hole in the spacecraft. We are not in danger, but the patch does not have any solar cells, so our solar power will be reduced somewhat. Maybe by 0.2% or so. This will mean extending our travel time by a few weeks due to less power for our nuclear engine."

Kevin ordered the rover compartment door to open, assessing the patch seal.

"No air is escaping; the patch is working."

"How is Candice?" Kevin asked Jennifer."

"Her arm is not broken, just bruised. I applied some salve. She will be fine."

"I don't know why the collision avoidance system did not work properly," said Kevin.

"Orbiter, can you enlighten us?"

"Yes, Kevin. The problem was that we encountered two Oort bodies. The collision avoidance system avoided the larger one, using the attitude control thrusters, but it could not deal with two at once. My software was not programmed for that contingency."

"Maybe I can fix that," offered Jason. "We have copies of both the source code and object code stored in our archives. I'm familiar with them from our training. Let me see if I can do something."

Later, Jason reported, "The fix is not difficult. We can run a second copy of the collision avoidance algorithm simultaneously. I can modify the operating system to enable this. I can verify the changes with our onboard simulator on the backup supercomputer."

"That's great," said Kevin. "Let me know when the testing is complete."

Two days later, "The fix is verified and validated, Commander. I'm ready to implement it. But we'll have to take Orbiter offline for a few seconds to upload it."

"We can afford a few seconds of downtime," said Kevin. "But is there any risk to the restart?"

"None. I know the system well."

"Okay. Go for it."

Ten seconds later Jason said, "It is complete."

"So now we can cope with two object threats simultaneously?"

"That's what the simulation indicates."

"Nice job, Jay. Once again, you know your stuff. I'm not sure I would be able to do it."

Almost five months later, Kevin received an audio from Mission Control. "Starship, we obtained an alert from your spacecraft of alarms from the cockpit. What happened? Is everyone okay? Is the spacecraft functioning?"

Kevin responded, although it would take another six months for Houston to receive his message.

"Yes, it was pretty exciting here for a while. We have a few bumps and bruises, but we are all fine. We collided with an object in the Oort cloud that tore a hole in the rover compartment of the spacecraft. We were able to patch it and the patch is holding. It turns out the collision avoidance system was not able to manage two threatening objects simultaneously. So, we modified the software to deal with this eventuality, should it be necessary in the future."

More than a year later, Kevin received another communique from Houston, "Thank you, Commander. Sounds like you have everything under control. Congratulations. You are quite resourceful. Sorry we can't be of any help from this great distance.

"More news from the home front. Congress has become more progressive. The President signed their bill to allow illegals to vote. Also, they have again crafted a bill to severely restrict gun ownership and sale to deal with the increasing crime in our cities. The Republican minority claims this violates the 2nd Amendment, and it will not affect our crime problem unless we deal with the perpetrators and mental illness more strongly.

"They continue to look for answers to our risk of T-Bond default. As much as one-third of our able-bodied population is still not declaring income or paying taxes. This frustrates their attempt to raise taxes to solve the debt problem."

"Kevin," said Sam. "Do we have to listen to this? It's so depressing. I'm glad we are not there to hear this every day."

"Orbiter, continue on to our destination," commanded Kevin.

***

Jennifer was giving Samantha her annual checkup when she announced, "Sam, your heart murmur is no longer minor. It is now so pronounced I can hear it without my stethoscope. Are you experiencing any chest pain, shortness of breath, fatigue, faintness or light headedness?"

"No. I feel fine. What does the murmur mean?"

"The blood in your aorta is back tracking after it flows through your heart, causing the sound. The valve is not closing completely. This reduces blood flow to your body. Sam, we need to replace your aortic valve."

"Oh no! I need open heart surgery?"

"No. Fortunately I can perform the replacement valve arthroscopically. It is minimally invasive. It is called a TAVR for Transcatheter Aortic Valve Replacement. I go through your femeral artery inserting a catheter with a collapsed artificial heart valve through your arterial vessel up through your leg and chest cavity into your heart. The valve is released into your existing valve and then is opened by inflating a balloon at the tip of the catheter, thereby replacing the function performed by the defective one without any invasive open-heart surgery. Then I extract the catheter. You will be back to normal within 24 hours."

"But where do we find an artificial heart valve?"

"Remember when I detected the small murmur in the first physical examination that I gave you back on Earth? I brought along two artificial Edwards valves made from cow heart tissue in case you or anyone else needed it. It is such a simple procedure yet can save a life."

"Will I have to take medication to keep my body from rejecting the cow tissue?"

"No. Your body will accept the artificial valve without immunosuppression medication. The vital parts of the body such as the brain and the heart enjoy immune privilege because they provide vital regenerative capacity."

"Jen, you are amazing. When can you perform it?"

"There is no immediate urgency, but I see no reason to postpone it. Your arteries are healthy, you have no infections, your mitral ring is strong enough to support the new valve, and you are not taking any blood thinners. Also, your creatinine level is a healthy 0.8, so the dye that I inject for the procedure will have no effect on your kidneys. If you are ready, we can do an ultrasound today. If that looks good, we could do it any day afterward. I will monitor you for 24 hours after the procedure. After that you will be as good as new."

"How long with the valve last? Could it wear out?"

"We don't have data to predict its longevity. But in the twenty years we have been performing this procedure, no artificial valve failures have been detected."

"I need to tell Mike."

"Of course, and if we proceed, we'll let the crew know. I'll ask Candy to assist. Don't worry, Sam. This was done at Johns Hopkins hundreds of times."

"Successfully?"

"Very successfully."

"Sounds like a miracle."

"Twenty years ago, this was only performed via open-heart surgery. Then it was a serious operation. We have come a long way with arthroscopic surgery."

The ultrasound results cleared the way for the TAVR procedure. Candice was happy to assist her mentor. Mike insisted on looking on. Jennifer prepared a sterile environment in Samantha's and Michael's quarters. She inserted an intravenous line into Sam's left arm and injected the sedation for comfort. It put Sam to sleep. She showed Candice how to use the X-ray machine to aid Jen in guiding the catheter containing the new spring-loaded collapsed valve through Sam's heart to the valve between the aorta and the left ventricle. She then remotely released the artificial valve, and it dutifully sprang into place inside the old valve. She then carefully extracted the catheter.

Sam woke up less than an hour later, feeling perfect. Her murmur disappeared.

"Wow, Jen," said Sam. "You are amazing. I feel like nothing happened."

"Trust me," commented Jen. "Something did happen. You now have a new aortic valve that is likely to serve you the rest of your life."

Candy and Mike were at least as impressed as Sam was.

***

"Kev," asked Jennifer, "do you think intelligent life exists outside the Solar System?"

"Well, actually I do. Think about it, Jen. Our own galaxy, the Milky Way, has over four hundred billion stars. They think at least 50% of these have planets. On average, the standard star has two planets in the habitable zone of a star where liquid water, crucial for life, can exist. So, there are likely hundreds of billions of Earth-like planets in our galaxy. While liquid water is a necessary condition, many others are required for all life, not just intelligent life. Based on our Earth experience, we need an atmosphere with oxygen to breathe, we need reasonably mild temperatures, we need food, we need energy, and we need to be protected from harmful solar radiation. These are just the basics, but they may be modified to sustain life. If all these conditions exist, then the issue is evolution.

"Scientists believe that life in some form on Earth appeared 3.5 billion years ago. But intelligent life on Earth began no more than 300,000 years ago. So, it took billions of years of evolution to evolve intelligent life. If we assume evolution occurred similarly on other planets in the habitable zone, only planets older than four billion years old would qualify. But the Milky Way is more than thirteen billion years old. So, there are plenty of planets with the right conditions for life that must have had more than sufficient time to evolve intelligent life. And that is only considering our galaxy. There are at least two trillion galaxies in the Universe.

"So, yes, I believe intelligent life exists outside our Solar System. Moreover, there must be intelligent life that has evolved much longer than we on Earth. So, I believe they are much more advanced than we are. To believe we are the most advanced form of intelligent life would be arrogant to the max."

"That is mind boggling, Kevin," said Jen. "So, if they exist and are so advanced, why haven't we encountered them?"

"Well, Robert Weryk discovered the first interstellar object passing through the Solar System in 2017. They named it Oumuamua, and it was so strange that a few scientists, including Avi Loeb, believed it was not natural but may have been a product of alien technology. It was cigar-shaped, estimated to be between 300 and 3,000 feet long and between 115 and 548 feet wide. It tumbled end-over-end rather than spinning. It could not be a comet because it had no visible coma. Some kind of force caused it to deviate from the natural hyperbolic orbit around the Sun. So maybe, we have been visited by aliens.

"But Earth is just a tiny speck of dust in this vast galaxy. Don't forget the Milky Way is 100,000 light years in diameter. That is almost six hundred quadrillion miles. So, maybe aliens don't know we exist. Or maybe they don't care because we are so backward."

"Wow, that is so humbling," declared Jen.

***

"Hey, check this out, Skipper," said Candice excitedly as she peered through the cupola. "Are those two bright stars Suns? They look like Earth's Sun did when we were more than one light year away."

"Yes, answered Kevin. "They are Alpha Centauri A and B, the two bright stars just 4.37 light years from Earth. They average only twenty million miles from each other and orbit around a common barycenter. Although it is too faint to see from here, our objective Sun, Proxima Centauri is near these Suns, only one-fifth light year from them. Imagine what our sky will look like once we land on our new planet, Proxima b."

***

This time it was Jennifer's turn to experience her own daydream while working out on the Peloton. Her mind drifted back to the anxiety and pride she acquired in the OR at John's Hopkins hospital. Although she was only a resident, she was given the honor to assist the chief of surgery in a delicate operation. Female conjoined twins were delivered by Caesarian birth just two days earlier. Jen had assisted in the birth which in part was why she was invited to assist in the risky surgery to separate the Siamese twin girls. They were fused at the back of their skulls and the lower backs of their bodies. They shared the occipital bone and were conjoined at the L7 bone and the buttocks. Fortunately, they did not share internal organs, enabling the separation surgery to save both of their lives. Thanks to the extraordinary skill of the chief surgeon, the operation was a complete success. Both girls survived – a rarity in such cases. Jennifer was beside herself with both relief and pride. But she also realized that she was fortunate. If one or both had not survived, she would be compelled to never agree to take part in such risky surgery again.

She was happy to end her reverie as she completed her hour-long exercise stint, sweating profusely not just because of the physical exertion.

***

A short blast from the collision avoidance system made everyone jump.

“Orbiter, what was that?” shouted Kevin.

“I don’t know commander.”

“Was it a comet?”

“No, sir. It had to be much smaller than a comet.”

Suddenly another alarm blast ran through their compartment.

“Wow, another one?” questioned Kevin.

“Yes, sir. But again, it is small. However, either it was very slow or very fast since the encounter was so brief.”

“Well, we are traveling at more than 70% the speed of light, so it must have been traveling slowly as we sped past it.”

Kevin joined Michael up in the cupola to see if there would be another encounter that he could observe. Suddenly Kevin thought he saw a black object much smaller than Starship slowly creep within sight on the starboard flank of their spacecraft. The collision avoidance alarm blared continuously.

"What the hell," blurted Kevin. "I don't know what it is, but it is keeping up with our speed. It is station-keeping with us!"

"Could it be one of those comet outbursts?" asked Jason. "Comet 17P/Holmes let out a huge flash of gas and dust in 2007 which looked like a flying hourglass."

"I think there's another one on our port flank!" yelled Mike.

Suddenly, as if a light were turned on, the black objects became bright and shiny.

"They look like flying saucers," offered Kevin. "Smooth, metallic-looking body, no visible engine.

They either are being piloted remotely or have intelligent pilots onboard."

"What do they want?" asked Mike. "Are we in danger of an attack?"

"I don't think so. It looks like they are monitoring us. But clearly, they want us to see them, or they wouldn't have lit up their spacecraft. The windows probably indicate they have some intelligent beings onboard, but I can't see anything through those windows."

"Neither can I," added Mike. "What should we do?"

"Nothing. Just calm down and let them examine us. They are not threatening us. Perhaps they are just curious."

Slowly, both alien spacecraft began barrel-rolling around Starship in tandem formation, as if inspecting it. Then Kevin and Mike felt a strange tingling on their bodies and a pop in their eardrums. After twenty minutes, the two alien spacecraft changed from a silver metallic glow to their black color. After station-keeping for over an hour, one of the disks sped off, no longer to be seen. But the other lingered alongside for another hour before rapidly departing. On its way out it seemed to dip left and right as if to say, "Goodbye for now."

The collision avoidance alarm stopped blaring.

"How fast were they going, Orbiter?" asked Kevin.

"At least 20% faster than we are," answered Orbiter.

Kevin and Michael climbed down from the cupola, rejoining the rest of the crew.

"What's going on, Kevin?" asked Jennifer.

"You are not going to believe it. Two disk-like spacecraft **actually rendezvoused** with us. Then they performed a precision station-keeping formation, one on each side of Starship. And we are traveling at 470 MILLION MILES PER HOUR! They wanted us to see them. Then they performed a roll maneuver, inspecting Starship as they rolled around us. Their windows indicated they were staffed by intelligent pilots. But we were unable to see through their windows."

"So, you think they were aliens piloting flying spacecraft?" asked Jenifer.

"I don't know what else to conclude," responded Kevin.

"They had to be aliens in flying saucers!" exclaimed Mike.

"Kev, we all experienced a tingling sensation while the alarm was blaring," offered Jen. "Do you know what caused it?"

"We had the same sensation," answered Kevin. "It is possible that they caused it, somehow. Also, we experienced a ringing in our ears. Did you?"

"Yes, we all did," answered Jason and Candice in unison. Sam and Jen nodded. "What do you think caused it?"

"I have a theory that these beings are intelligent with advanced sensory perception. They were not hostile or threatening us but were curious and examining us. I think they used some kind of radiation on us to see what makes us tick."

"Really? Responded Jason. "That's pretty wild. Do you think they will continue to track us?"

"Perhaps."

"What do you think they want?"

"I have no idea. We may have entered their domain and want to make sure that we are not threatening. But I can only hope that their apparent lack of hostility means they may be friendly. If not, we would be in big trouble. Their spacecraft propulsion, navigation, and control are more advanced than anything I've ever imagined. We may be in for an adventure the rest of our journey."

"I pray you are right about their intentions," said Mike.

"Well," said Jason, "I thought after leaving the Solar System we might get bored. This is pretty darned exciting."

"Do you think we can learn anything about these creatures, Kevin?" asked Jennifer.

"If my theory is correct, I doubt it. If they learned all they need about us, they may leave us alone. I'll report back to NASA and see if they have any ideas, but it will take over four years to get an answer."

"I won't be able to sleep tonight," said Sam.

"Join the club," responded Mike.

"Great material for my journal," exclaimed Candice excitedly.

Kevin reported their unbelievable encounter on its two-year transmission to Earth. He hoped NASA would not consider him unstable and delusional. They might ask for evidence he could not provide.

"Houston, you are not going to believe this, but we just had a close encounter of the second kind. Two large shiny spacecraft rendezvoused with us while we were traveling at 70% the speed of light. One stayed with us at approximately one hundred feet from our cupola for an hour, then sped off at an incredible speed. The other stayed with us another hour. We all had strange tingling sensations and our ears were ringing during the encounter. Can you help us to understand any of this?"

Thirty-six hours later, Sam admitted to Mike that she was having nightmares.

"What about?"

"Interplanetary aliens."

"Sounds spooky, but I can understand after that close encounter with the saucers. But they have left us alone, so there is probably nothing to worry about."

"But my dreams are frightening."

"My solution is to hold you tight tonight," responded Mike, smiling.

***

Kevin announced that they had reached the halfway point in their journey. He shut down the ion engine re-enacting zero G freefall.

"Sorry team. I had to shut down our ion engine to coast to our next waypoint. We'll be in zero G for the next few weeks to avoid overshooting our waypoint. We are currently traveling at nearly 90% the speed of light. The relativistic effect results in our increased mass. This has been causing our acceleration to decrease because our nuclear engine thrust is a constant. But we are on schedule. We have traveled more than two light years. We can soon begin our steady braking maneuver for the rest of our trip. We have to reduce our speed to enable capture by our target star."

The crew actually enjoyed their weightlessness for the interval. Three weeks later, "Orbiter, turn Starship 180 degrees and continue to navigate to the Proxima Centauri star." Kevin restarted the ion engine to initiate braking.

"Will we see any planets or moons as we enter Proxima's sphere of influence?" asked Jason.

"We don't know," answered Kevin. "Two other planets were discovered in 2019: Proxima Centauri c and Proxima Centauri d. They are not in the habitable zone because they are too far away.

"As we learned in training, their star, Proxima Centauri, is a red dwarf with a mass only one-eighth that of the Sun and one-seventh the Sun's diameter, but thirty-three times the Sun's density. Our Proxima Centauri b's orbit is only 4.5 million miles in radius and has a period of 270 hours. But Proxima Centauri c's orbit is 140 million miles in radius; too cold to be habitable, and a period of over five years. Closest to the star is Proxima Centauri d at 2.6 million miles with an orbital period of only five days, too hot for habitability. So, it is possible to fly by Proxima c, but not d.

"Obviously, we are entering the unknown and unexplored. So, who knows, maybe we can see planets and moons unknown to Earth's astronomers.

"But first we should be able to see the binary stars, Alpha Centauri A and B before we see our dim Proxima Centauri. If you remember, we touched on all this in training."

The next week, Kevin was in the cupola looking for Alpha Centauri. Instead, he felt a jolt as the alarm system blared.

"They're baaaaaack," he announced, mimicking the Poltergeist movie.

Everyone quickly crowded into the cupola. They observed the two spacecraft matching their speed on either side of Starship. The beings, whoever they were, turned on the light that revealed their spacecraft's metallic bodies.

"This is so frightening," said Samantha.

"So exciting," said Jason.

"Unbelievable," said Jennifer.

"Kevin, what do they want?"

"I think *(I hope) they* just are curious. I don't know if these are the same guys who looked us over earlier, or their colleagues who want to check us out firsthand."

Once again each felt the tingling that they first experienced.

"I think they are probing us with some kind of radiation. Either exploring our bodies or our brains or both," added Kevin.

Suddenly both spacecraft flashed their lights on and off.

"That could be a "hello" signal to calm us and let us know they are friendly. Make sure you are all thinking kindly thoughts."

They continued to experience the tingling.

"I can't take this," said Sam exiting the cupola.

The saucers continued their tight formation for almost one hour before turning off their lights and speeding away.

"Well Candy, this is great stuff for your journal," quipped Jason.

"I'd give anything to know about these aliens," said Mike.

"Well, maybe we will get the chance," said Kevin. "I'll bet we will have another encounter, now that they examined us and know we are not hostile. I wonder if we are intruding on their turf."

***

When Mike completed his hour-long turn on the Peloton, he felt lightheaded and had to lean against the Starship wall to steady himself.

"*That is strange,*" he thought. "*What is going on?*"

The next day, he had a similar occurrence as he got off the Peloton.

"*This is crazy!*"

Samantha noticed his instability. "Are you okay, Mike?"

"Of course. I am fine," he answered, a bit embarrassed.

"Well, you look a little weak."

"Weak? You must be kidding. I'm as healthy as a horse. Just a little tired. I probably didn't get enough sleep last night. And that was your fault, hot stuff! I am an athletic specimen. I was the top linebacker in high school and haven't lost a step."

"Okay, okay," answered Sam, smiling. I'll take it easy on you tonight."

The next day Sam followed Mike into the gym again to observe. "Mike, you're listing to the left. And shuffling your feet. Something is wrong. You have to have Jen examine you."

"Well, okay. If that will satisfy your motherly concern," retorted Mike.

They approached Jennifer together.

"Jen, Mike is having balance problems. Could you check him out?"

"I keep telling her, I'm okay, but her motherly instinct insisted I get your opinion," said Mike.

"Okay, Mike. Walk up and down this short path," said Jen.

"Well, you are shuffling your feet and look a bit off balance," observed Jennifer. "Please hold your hands steadily out in front of you." Jennifer watched his fingers closely. "Now rapidly touch each finger to each thumb."

"Mike, you have a slight tremor. How long have you been experiencing balance issues?"

"I'm okay. But when I get tired, I do feel like I lean to my left side. Do you think there is something wrong with me?"

"Mike, you have early-onset Parkinson's Disease," diagnosed Jennifer.

"Are you sure?" he asked with trepidation.

"One hundred percent," she answered confidently, but with empathy.

"Wow!" he gulped. "What does that mean?"

"Parkinson's is a long-term degenerative disease of the central nervous system. It is manifested in a shortage of dopamine in your brain. Millions of people, particularly men, have PD. It is unusual for you to acquire it at such an early age, but it happens. The actor, Michael J. Fox, was diagnosed with it at age 30, but didn't manifest obvious symptoms until a few years later. You will not die from it because you are otherwise healthy. But as the years go on, your bodily motor skills and perhaps your cognizant acuity will decline."

"How could I even contract this terrible disease?"

"The medical community does not really know. There is a possibility that genetics and some environmental factors may play a part. It is extremely unusual to manifest the symptoms at your age. Have you ever had a head injury or been subjected to pesticides?"

"No pesticides that I can recall, but I did have a concussion in high school playing football. Did that cause it?"

"They may be risk factors. Again, we're not sure, but that could be a factor."

"Is there any cure?" asked Sam.

"I'm afraid we have no definitive answers as yet. However, in 2022 the medical department at University of California-Irvine started a clinical trial program. They implanted dopamine-producing stem cells into the part of the brain that regulates movement for patients that qualify. Then the patients took medication to partially suppress the immune system to prevent rejection by the body. But it will take years to establish the safety and efficacy of the procedure and whether the cells survive, and motor symptoms improve.

" For now, Mike, I can give you a strong medication to reduce the symptoms. It is called Carbidopa-Levodopa. You will start with just twenty-five milligrams per day, gradually increasing to three times per day. The stuff is powerful, so begin taking it with food. Later, if your system tolerates it, you can ingest the pill without food. I'll monitor the effects."

"Is it contagious?" asked Mike with fear in his voice.

"No. Not at all."

"This is pretty hard to take, Doc," said Mike.

"I'm sorry."

"Please don't reveal this to my teammates."

"Why not?" asked Sam.

"I don't want them to think I am defective in any way. I need to be a fully functioning member of the team. Also, I don't want anyone's pity. Both of you, please promise me that you will keep this to yourselves."

"You have my word as a doctor not to disclose your medical condition without your permission," responded Jennifer.

"I promise, darling. You can trust me," added Sam.

"Mike, are you having any trouble sleeping? Any nightmares?"

"Well, now that you mention it, I have not been sleeping longer than a few hours before I wake up. Then I have trouble falling asleep again. No nightmares, but some extremely vivid dreams that are sort of entertaining."

"Those are some of the symptoms of Parkinson's," said Jennifer. "I'll give you a supply of Melatonin. Take five milligrams one half hour before going to bed. It should help you sleep."

After two days of ingesting the Levodopa, Mike was amazed at how his gait and balance had improved. He was grateful to Jennifer and her medicine. He now trusted her diagnosis completely and her prognosis.

His self-esteem was restored, at least partially.

***

The next night, Sam's dreams escalated. Mike found her flailing wildly.

"Bad dreams again, Sam?"

"They are getting worse. Aliens again."

"I have an idea, "said Mike. "Jason told us at his interview that his hobby was hypnosis. Maybe he could help."

"Worth a try," said Sam. "I have to do something. I'm not getting enough sleep, and I'm keeping you awake too. I think Jason is in the gym."

"Jason, Can hypnosis work for getting rid of nightmares?"

"Yes, I think so. You are having nightmares?"

"Yes. Ever since the flying saucer encounters, I have been dreaming about aliens."

"What do they look like?"

"They get worse each night. Now they are little green men with bug eyes, no ears, a small nose. and a mouth that displays pointed teeth in a hideous smile."

"Wow. If they really look like that, let's hope we never experience a "Close Encounter of the Third Kind.," said Jason, hiding a smile.

"As a child I used to think about aliens ever since I heard about flying saucers. I remember being so afraid of them when I was seven or eight."

"I want you to sit on this stool and relax. You will actually enjoy what I am about to do.

"Sam, relax, listen to my voice, and do exactly what I suggest. Trust me.

"Relax but keep your head immobile and let your eyes follow the tip of my index finger. Focus on my finger as I move it."

Jason moved his finger slowly, back and forth across Sam's vision.

"Your eyelids are getting very heavy. You only hear my voice. You can hardly keep your eyes open. You are becoming so tired. Close your eyes, Samantha. You are falling fast asleep. Deeper and deeper asleep. You are feeling so pleasant. You want to sleep but listen to my voice.

"Samantha, extend your left arm. Your arm is becoming rigid. You cannot bend your arm. Try. You cannot bend it."

"Okay, Samantha, You can now bend your arm. Relax. Just listen to my voice. We are going back in your life. You are now in primary school. You are eight years old. What grade are you in?"

"Third grade," responded Sam very slowly.

"Do you see alien creatures?"

"Yes. They are horrible."

"Samantha. These creatures are not horrible. They like you and you like them. They are friendly. They look like your classroom friends. Can you see them?"

"Yes."

"What do they look like?"

"My playmates."

"Do they want to hurt you?"

"No. They are my friends. They are friendly."

"Very good, Samantha. Now we are coming back to current time. You are growing up. You are an adult. Do you see any aliens?"

Sam fidgeted on her seat on the chair, frowning.

"Yes. They are horrible."

"Samantha, the aliens want to be your friends. They are pleasant looking. They have flesh colored skin; they look like you. They have beautiful eyes, perfectly formed ears and noses. Their smiles show pretty teeth. They want you to like them. They want to be your friends.

"Can you see them, Samantha?"

"Yes'"

"What do they look like?"

"They look like me. They are smiling. They want to be my friends."

"Can you be their friend?"

"Yes, They ARE my friends."

"Samantha, when you wake up, you will no longer dream of hideous aliens. Instead, you will see them looking like friendly, smiling humans.

"When I count backward from ten, you will awaken refreshed and rested.

"10, 9. 8, 7, 6, 5, 4, 3, 2, 1."

Sam opened her eyes.

"How do you feel?"

"Great, except my left arm hurts. When are you going to hypnotize me, Jason?"

"Sam, you are a perfect subject. I doubt if you will have any more nightmares."

"Really? It is that easy?"

"Yes."

Sam enjoyed her dream that night of seeing the human-looking aliens exiting from their spaceship. She was happy to meet her new friends. Mike finally got a good night's sleep.

9. Alpha Centauri

"When you get a chance, come up into the cupola and take a look at the binary star, Alpha Centauri," suggested Kevin.

"Yes," responded Jason. "I can see the separation between Alpha Centauri A and B. It no longer looks like a single bright star. One is much brighter than the other. I can't see our Proxima Centauri yet even though it is closer. I guess it is just too dim."

"That's right," continued Kevin. "Alpha Centauri A's official name is Rigil Kentaurus. From Earth, it is the third brightest star in the sky. The official name for

Alpha Centauri B is Toliman. They are 4.37 light years from Earth.

"Alpha Centauri A is 50% more luminous than the Sun even though it has almost the same mass. It is more than 20% bigger than the Sun and is less dense. Alpha Centauri B is 35% smaller with 10% less mass than the Sun and half its luminosity. Our Proxima Centauri is 1.2 billion miles from them, is 20,000 times less luminous, and only four times the size of our Moon. We will have to be a lot closer before we can see it.

"Houston," announced Kevin. "We have passed the half-way point and have begun the braking thrust. We have a good visual of the Alpha Centauri binary but cannot see Proxima Centauri as yet. By the time you receive this message, we should be in the sphere of influence of Proxima Centauri."

***

Two years later NASA Mission Control announced, "Starship, the asteroid Apophis just hit North America. It was ¼ mile wide and traveling fast enough at impact to produce the equivalent of twenty million tons of TNT. It completely destroyed the entire city of Stanstead, Quebec, Canada and most of the adjacent town of Derby Line, Vermont. The devastation rained over much the Northeast region of the United States and the Quebec Province.

"We had been tracking its trajectory for at least 15 years including its flyby in 2029. We tried to nudge it off its course by colliding it with our asteroid chaser, but the resulting change in its velocity was negligible. We did not try to blast it into smaller pieces with a nuclear warhead. Our simulations showed the impact would not affect the trajectory, and the remaining mass impact would not significantly change the result.

"More than 100,000 American and Canadian lives were lost. It wiped out animals and plants within four hundred miles of the epicenter. Billions of dollars in damage is anticipated. We will know more when FEMA is able to sift through the site.

"We were fortunate the asteroid was not larger nor struck a larger city. The catastrophic devastation could have been much worse."

Kevin said, "On top of all the mess being created by our fellow man, now we experience a natural disaster that once again the population is unable to do anything about."

"Is there any end to the bad news from Earth?" asked Jennifer.

The rest of the crew remained silent and sullen.

***

"Kevin, I think I see Proxima Centauri," said Jason. It is pretty dim and not too large."

One year later Kevin announced, "We are entering the sphere of influence of Proxima Centauri, gang. We are three billion miles from our new star. Our speed is down to ten million miles-per-hour. Since this is uncharted, we know little of what we may encounter. We know of three planets: in the solar system of Proxima Centauri: Proxima b, Proxima c, and Proxima d. We may encounter Proxima c first since it is the largest and farthest from Proxima Centauri. It is 150 million miles from the star. We believe it may have a ring system. We will do a flyby to get a negative gravity assist to help reduce our speed.

"Orbiter, fly within two hundred miles of Proxima c's leading edge. Time the flyby such that our velocity vector continues its path to our Proxima b destination."

"Commander, we will have to continue our ion engine braking for the next 2.5 billion miles, then fire our

Raptor engines for 14 seconds to slow us down enough to obtain the desired speed and alignment for the inverse gravity assist."

"Great. Do it."

After the burn, Kevin received a communication from Mission Control.

"Starship. Alert! We just learned from the James Web telescope that we have a problem. Proxima Centauri is emitting massive solar flares. These are transmitting billions of tons of plasma with ultraviolet radiation and x-rays out into space. This is not that unusual because our own Sun is subject to these episodes although not as dramatic. However, we learned that Proxima Centauri's superflares have recently become much more active and are bombarding our exoplanet unmercifully. As a result, these flares are rapidly depleting the atmosphere of Proxima b and may have been doing so for a billion years. This leaves the surface much more vulnerable to radiation. At this rate of depletion, the planet is in danger of eventually losing all its atmosphere. We had thought its atmosphere would protect the colony from intense radiation. As far as we know, the planet does not have a magnetosphere nor possess radiation belts like the Van Allen belts to deflect this radiation onslaught and protect the atmosphere from destruction.

"We no longer believe it is safe to live on Proxima b. You would have to fortify your habitat against radiation much more than the lunar colonies are.

"Commander, you need to reconsider whether to land. You should consider returning to Earth, rather than expending the propellant to land there. You can preserve it for landing back on Earth."

"Wow!" exclaimed Kevin. "You heard all that. We have a decision to make."

"Kevin, how could you have been so wrong?" asked Candice. "How could NASA have been so wrong?"

"Incompetent!" added Jason. "We've devoted eight years of our lives to this folly adventure. We should have waited until they had more data. I can't believe NASA could be so sloppy."

"I agree," answered Kevin. "I should have asked more questions; demanded more information; more research; not accepted their pronouncements at face value. This is my fault."

"Wait a minute," said Jennifer. "We all made our decisions individually and independently without coercion. Any one of us could have questioned the soundness and accuracy of the information before we made such a profound commitment. We had a year on Earth to do so. We can't lay this all on Kevin."

"In any event, we have a major decision to make -- accept NASA's latest data and recommendation and decide to go home, or to land and deal with the threat," said Kevin.

"We should also consider that NASA detected this onslaught of superflares at least eight years ago: four years for the observation of the event to reach Earth; another four years for their communication to reach us. We don't know whether the superflares have subsided or whether they have gotten worse in eight years."

"How much time do we have to make our decision?" questioned Michael.

"We have an indefinite amount of time to decide," declared Kevin. "We can burn some fuel to slow our speed and orbit the planet, examine the topography from space for a while, then decide whether to land or return to Earth."

"Another eight years?" complained Samantha. "Then return to the messed-up world we left? I don't think so."

"Sam is right," said Mike. "We should do whatever we can to carry out our mission, not for NASA's sake, but our own."

"I agree with Mike," said Jen. "We should see this through. Okay, so it is more risk than we bought into. But the alternative of returning to a dysfunctional society is not for me."

"Well said, Jen," agreed Candice. "Even with the added risk, we have the opportunity to create a much better community than today's America will allow. And EIGHT MORE YEARS of travel? That's not for me."

Jason concurred. "When you really consider the alternative, the choice is clear. Had we even known about the risks early enough, I would have chosen to go, especially after hearing the news of the screwups on Earth."

"So, it is unanimous?" asked Kevin. "We proceed and reject the idea of returning to Earth?"

Everyone agreed. Kevin was pleased and proud of his crew.

"I'm happy we reached this decision without considering one more issue that occurred to me. We could have a risk of entering Earth's thick atmosphere. The patch we used to repair the hole caused by the collision with the Oort cloud object might not survive reentry. Remember we lost the Thermal Protection System tiles in the vicinity of the hole. The tremendous heat could cause the special adhesive on the patch to fail. If it did, the hole could reopen with disastrous results because the inner thermal blanket may not survive without the tiles protecting it. But this should not be as much a problem in the thin atmosphere of Proxima b."

"On to Proxima Centauri c to use its gravity to slow us down and then on to b, Orbiter."

"Houston, we heard you loud and clear. But we have decided to accept the risk and intend to land on Proxima b as originally planned. We will keep you abreast of our progress. Over and out."

***

"Crewmates, Michael and I are happy to announce that we are pregnant," Sam proudly proclaimed. "Jen confirmed it yesterday. But the baby won't be born for seven months, well after we land."

"Wow, it's happening already," said Kevin.

"I'm so happy for you," responded Candy. "The colony's first natural born citizen! Jason and I hope to follow soon."

"And we as well," added Jen. Kevin nodded, smiling.

"We've started making baby clothes and diapers from the cloth NASA provided. We expect to do the same for our own new clothes after we land."

"Do you know the sex? Have you chosen a name yet?" asked Candice.

"No, we want to be surprised. Only Jen knows the gender, and she has promised not to tell."

***

Forty-five days later. "Commander," said Orbiter, "we are now twenty-five trillion miles from Earth. We are two million miles from Proxima Centauri c. The planet is 138 million miles from its star, Proxima Centauri. Its orbital period is 1928 days. It is seven times the mass of Earth, therefore classified as a super Earth even though its diameter is smaller. Due to its vast distance from Proxima Centauri, it is uninhabitable, with a temperature of 390 degrees below zero. We will soon see its ring system. Its orbit is inclined to the plane of Proxima Centauri's rotation by 133 degrees."

"Thank you, Orbiter. So, this is the first planet of our own new solar system," declared Kevin. "Its large mass and speed should decrease our speed significantly relative to its Sun as we execute a reverse gravity assist around its forward edge."

Fifteen minutes later, Mike had climbed into the cupola and announced, "Kevin, I think I see its rings." The others joined him to view the two spectacular rings of orange and blue circulating the planet some 450,000 miles away.

"What a sight! So vivid and beautiful," exclaimed Candice. "This is even more impressive than Saturn's rings. What an entry for my journal."

"Unbelievable. It lights up the planet," echoed Samantha.

"I wonder what sights the rest of this solar system has to offer," said Jennifer.

Minutes later, Jason said, "Hey, look. Is that a moon?"

"I bet it is," said Kevin. "Smaller than Earth's Moon, but interesting. Orbiter, will we get closer to it?"

"Our trajectory should take us within 1,500 miles."

That got the crew even more excited.

"It is pockmarked with craters just like Earth's Moon," said Mike. "But it is a brown, almost reddish, color."

"Right," said Jason. "That must mean iron covers its surface. Let's use our spectrometer to tell us the chemical composition."

Jason trained the spectrometer on the moon. "Wow! The spectrometer indicates a treasure trove of chemicals. In addition to iron, it shows sodium, magnesium, silicon, titanium, chromium, and cobalt. A mining goldmine. We're not in Kansas anymore, Toto."

"Now we can see the planet and its features," said Jennifer. "This is so exciting!"

"It has mountains and valleys, but I don't see any water," said Candice.

"Right." Said Kevin. "It is too cold. But I think I see ice caps at both poles."

"It has enormous plumes that may come from volcanic eruptions," observed Mike.

"Yes," said Jason. "They told us it may be seismically active."

"Approaching periapsis," announced Orbiter.

"Hey, there are two more moons," exclaimed Jason. "They look smaller than the other one."

"What fun," said Sam. "I feel I am in a video game."

After the periapsis flyby, Spaceship coasted outbound from the planet, out through the beautiful rings.

"Goodbye interesting planet," said Candice, smiling. "Sorry we have to go. Maybe we'll be back someday on holiday."

"Okay Orbiter. On to Proxima b to find us a home," said Kevin.

***

Kevin called an important planning meeting to discuss plans for their upcoming home.

"There are many facets of our lives on the new planet which will require adjustments as we learn about our new environment. Still, it is important to agree on some fundamentals for life in our new community which we can agree upon before we land. First, how should we govern?"

"Do we really need any government for just the six of us?" asked Candice. "One benefit of leaving Earth is we no longer have someone else to control our lives."

"Perhaps not right away," answered Kevin. "We proved our great relationships, developed over the past eight years demonstrate our mutual respect and compatibility. But we will soon be raising three families and eventually our little community will grow into a colony. So, I believe establishing some rules of governance will be of benefit to resolve any conflicts in the future."

"So, what do you think we need to establish?" asked Jason.

"For lack of a better word, some laws or rules to ensure our individual rights are not infringed upon. So first, what are those rights?

"We don't have to derive them from whole cloth. How about starting with the United States' Bill of Rights, the first ten amendments to the US Constitution? It was originally based on the Ten Commandments and the Magna Carta which served England well since 1689 and France since 1789. It certainly worked for the American colonists."

"But the American colonists had hundreds of people with diverse backgrounds to address," said Jennifer. "Will that apply to us?"

"Well, let's look at that and consider the eventual growth of our colony. The intent is to guarantee preservation of those freedoms' rights and limit the government's power over its constituents."

"This makes a lot of sense to me," said Michael. "Let's vote on it."

"Wait a minute," interjected Jason. "Are we a colony of the United States, or are we independent?"

"Good question," said Kevin. "In the traditional sense, we are a colony of the United States. The US government funded us, equipped us, and supported us. On the other hand, we are four light years away. Our round-trip communications take more than eight years. There may be nothing America can do for us or us for it. So, in a practical sense we truly are independent. If the US government makes decisions for its people, do we abide by those decisions? Let's discuss this."

Samantha was the first to respond. "I strongly believe we should become an independent colony. This is one of the reasons I wanted to be part of it. We have the opportunity to free ourselves from the terrible slide toward socialism that is overtaking America today. I am a strong believer in the strength of humanity to be self-sufficient and productive; not to be tempted by government handouts. We can do this if we are independent of the whims of a progressive government."

"I agree totally," chimed in Jason. "Self-reliance without government intervention is what made America great. We need to return to those values. I don't think we can do that if we have to adopt what comes from the US administration and Congress."

"You both make a lot of sense," said Michael. "I'm convinced. I vote for the independence of our colony."

Jennifer agreed. "Wow! I couldn't have said it better. Your strong beliefs resonate with my own."

Candice jumped in, "It should be no surprise where I stand on this. The US government has become weak and feckless in foreign policy. And the local governments are soft on crime, causing the cities to crumble. I fear for the future of the United States if the politicians don't turn this around completely. I don't want to be governed by the people in charge of these disasters. I vote for independence."

"I can't disagree with your views and assessment of our political issues," responded Kevin. "But consider what NASA and the country has done for us. And all of us have thrived in our careers despite the recent political environment. Frankly, I am torn on this question."

"Well, the six of us are a democracy," said Jason. "So, let's vote on it."

Kevin took the lead, "Any more debate? Okay. All in favor of declaring that we are forming a colony independent of political ties to the government of the United States, raise your hand."

All five hands immediately shot up. Kevin slowly raised his hand as well.

"Do we have to notify America with a Declaration of Independence as the original colonists did with England in 1776?" asked Mike.

"I don't believe we have to inform America of this decision, at least not now," suggested Kevin. "It would certainly offend NASA as well as the US citizenry. Until we incur a government conflict, which may never happen, we can stay silent on the issue, and avoid an unnecessary confrontation."

The crew agreed.

"Candice, please carefully document all of this in your journal," requested Kevin. "We should all review and sign it."

"Kevin, does this mean we have renounced our US citizenship?" asked Samantha.

"Not at all. We can be citizens of the new colony and still be US citizens. Dual citizenship."

"Back to the question of adopting the US Constitution's Bill of Rights," interjected Michael. "Are we ready to vote?"

"First, let's study them individually before voting," suggested Sam.

Kevin passed around his copy of the Constitution.

"Any comments, suggestions?"

"The US forefathers were smart people," offered Candice. "Let's vote."

"All in favor of the US Bill of Rights, raise your hand," said Kevin. "It is adopted unanimously."

"Kevin, any other issues to address for our new colony before we land?" asked Jen.

"Well, we are all going to be busy learning about our unique environment and establishing our residence.

Michael can take the lead in constructing our home. It will require all of us to follow his lead. Sam will set up her garden. Jen and Candice will establish their medical facilities. Jason and I will help wherever we are needed. And I will provide whatever management is required."

"Wait a minute," interjected Candice. "You are the commander so long as we are aboard Starship, but once we land, that is no longer true. As a democracy, we must elect a leader or mayor of our new community."

"You're right," responded Kevin. "Let's elect a mayor to provide the leadership required for our new community."

"Well, speaking for me, I can't think of a better person as our leader than the one we have been following for the past eight years--actually for over nine years," offered Jason.

"Although I am obviously prejudiced, I totally agree," responded Jennifer. "Kevin should continue as our leader and mayor."

"I think we all have leadership talent," said Candice. "But I will be happy to continue following Kevin's lead."

"Me too," agreed Michael.

"I emphatically agree," chimed in Samantha. "Let's vote."

Kevin was unanimously elected as mayor of the colony.

"Thank you for your confidence," responded Kevin. "I will do my best to live up to your expectations."

Michael suggested, "We need a name for our colony and a name for our new planet. I don't think 'exoplanet' or 'Proxima Centauri b' does it justice."

"Right," responded Kevin. "First, let's name the planet. Any ideas?"

"How about the New World?" offered Samantha.

"How about continuing the names of the solar planets, named after Roman Gods?" suggested Jason. "Maybe Apollo or Vulcan?"

"Or Cupid," suggested Candice.

"I think we should use the names of Greek gods, analogous to the Roman gods," said Jennifer. "Maybe Aphrodite, Athena, Artemis, Hercules, Zeus or Hermes."

"That's a great idea," echoed Kevin. "That way we can name the rest of the planets in this solar system just like Earth's Solar System but using Greek instead of Roman names."

"I'd prefer a female god," said Candice.

"I like Zeus," offered Samantha. "The boss of all the gods."

"Nice," said Kevin. "That would give our planet the stature it deserves."

"Works for me," said Michael.

"Me too," agreed Candice. "Let's vote."

"All in favor of naming our new planet Zeus?" asked Kevin.

Once again, it was unanimous.

"Okay, now our colony."

"Maybe we should wait until we really experience the environment," suggested Jennifer.

"Let's tentatively pick a name, then see how it fits when we arrive," offered Kevin. "Any ideas?"

"How about "Shangri-La?" suggested Sam.

"I hope it will appear that hospitable," said Jason, "but I won't hold my breath."

"How about Newland?" offered Michael.

"I like it," said Jen. "It fits no matter what we find."

"Me too," agreed Candice.

"How about Alien Place?" suggested Sam. "No, strike that. Too foreboding."

"Any other ideas?" questioned Kevin.

"Okay, let's vote. "All in favor of calling our new community Newland?"

Reluctantly, everyone agreed with Newland.

"Wait a minute," said Jen. Let's not get ahead of ourselves until we see our new digs."

"Maybe we will reconsider Shangri-La and Alien Place once we learn exactly what we're getting into," said Jason.

"You are right," said Jennifer. "Newland sounds unexciting and bland."

"Kevin, if we have established laws for our community, won't we need an executive body to implement the laws?" asked Jason. "And then a judicial system to enforce them?"

"Maybe eventually, but we'll have plenty of time to decide that."

"Eventually as we have children, we will need a solid education system," said Michael. "But again, this can wait."

"Wow, so much to consider," said Sam. "Creating a new community is not easy."

"Kevin, how did the lunar colonists manage these issues?" asked Jen.

"First, they are not independent, but are an official colony of the United States, even though the citizens of several countries populate them. This made all the rest of the decisions easier. So, they adopted the Constitution and the three branches of government: executive, legislative, and judicial. Of course, their infrastructure to implement these branches is minimal, since they only had twenty-one residents at the time I was there. Also, their government challenges, driven by US structure, were minimal since they had to be unified in their objectives and values. Finally, they had almost continual communication with Earth since the time delay was only 1.3 seconds.

"While some of our challenges are similar, others are quite different. Still, we can learn much from them in building our community."

"I'm happy we have you to guide us," offered Candice.

***

"Commander, we are experiencing a small force trying to pull us off our planned trajectory toward Proxima b," said Orbiter.

"Strange." said Kevin. "That could only be a gravitational force. But we are not aware of any massive objects in this vicinity."

"Maybe it's another planet," offered Mike.

"We don't know of any other planets. But that makes sense. Can you see anything?"

"Not yet."

"Orbiter, let's investigate."

Orbiter complied with a thirty-eight second burn of the Raptor engines to change the trajectory toward the gravitational force.

"There IS a planet," said Mike. "I see it!"

Everyone rushed into the cupola.

"Cool," said Candice. "And no one has ever seen it, even with telescopes?"

"Right," said Kevin. "We have discovered a new planet."

"Can we see it up close?" asked Jason.

"You bet," answered Kevin, excitedly. "We have plenty of fuel.

"Orbiter, how much fuel would it take to orbit this planet?"

"We only have to slow down to 4,500 miles per hour. It will take a burn of 65 seconds to reach orbit if we pass in front of the leading edge. We still would have enough reserve to go on to Proxima b and land."

"Okay, Orbiter. Let's do it."

***

Orbiter announced, "Commander, we are now entering the planet's sphere of influence at 400,000 miles from its center. Its mass appears to be slightly less than the mass of Earth."

The entire crew crowded into the cupola.

"From here, It appears to be about the size of Earth," observed Kevin, recalling his vision from his 30 days on the Moon.

"And it looks like Earth: a blue dot," said Mike.

"This is so exciting," said Candice.

"Kevin, could this planet be an alternative to Proxima b?" asked Jennifer.

"Well, I think it may be in Proxima Centauri's habitable zone, but let's not get ahead of ourselves."

"I think I see clouds!" exclaimed Samantha.

"I do too," agreed Mike.

"That's a great sign," said Kevin. "It must have an atmosphere."

"And clouds mean water vapor," added Sam.

"Jason," said Kevin. "Use the spectrometer."

"Hey, it not only has real water vapor, but nitrogen, methane, hydrogen, and OXYGEN! I can't measure how much, but it has to be more than just a trace."

That got everyone excited.

"Commander, we are approaching periapsis from the leading edge of the planet," announced Orbiter. "Do you want me to proceed with the orbital burn?"

"Absolutely."

Kevin shut down the nuclear engine. Starship was back in a zero G environment until the Raptors ignited. After the forty-five second burn, Starship was in a 200-mile circular orbit with a period of 86 minutes. Once again, they were on a zero G coast. They had a tough time staying up in the cupola without bumping into each other. But nothing was going to keep them from pressing their noses against the viewing bubble.

"It DOES look like Earth," said Candice. "It is beautiful."

"I see oceans and maybe rivers," said Jen.

"Don't set your expectations too high," said Kevin. "That could be more liquid methane than real water, as could the atmospheric vapor."

"Kevin, we have to land and find out," said Jason.

"Orbiter, do we have enough fuel to land, then lift off, go on to Proxima b, and land there?"

"No, Commander, we could lift off, then orbit at Proxima b, but we would run out of fuel during the descent to Proxima b."

"What if we don't orbit Proxima b, but make a direct descent?"

"That may not save enough fuel to make it, Commander."

"Okay, crew. We have another critical decision to make."

"Why doesn't NASA know about this planet to give us information about its atmosphere and informed advice?" asked Mike.

"My guess," said Kevin, "is the astronomers never discovered this planet because its orbit is inclined to its ecliptic plane. So, it does not transit Proxima Centauri along the line of sight from Earth. Also, its smaller mass and distance from the star does not cause the wobble that Centauri b does, minimizing the Doppler effect on the planet."

"So, we're on our own," said Mike.

"And I see the auroras, one near each pole," said Jason. "That means this planet may have a magnetosphere."

"Well, there are four conditions for a planet to have a magnetic field," said Kevin. "The planet must rotate at a high enough speed; it must have a liquid core; the interior fluid must have the ability to conduct electricity; and the core must have an internal source of energy that propels convection currents in the liquid interior."

"Let's hope it has all four," said Mike. "Then the surface would be protected from solar flares just as Earth is."

"I think we should land here and take our chances," said Jason.

"I agree," said Candice.

"The radiation risk would have to be less than Proxima b," said Mike. "The heavier atmosphere should offer more protection."

"Right." said Sam "And we would be much farther from the Proxima Centauri superflares."

"Well, I believe we are in the habitable zone," said Kevin. "So perhaps the oceans and rivers we see are water after all."

"So, let's do it," said Jennifer.

"Unanimous again," said Kevin. "What a great crew.

"But first we have to select a landing site."

"Orbiter, reduce our speed a bit to reduce our periapsis to 85 miles."

Orbiter popped the Raptors for a three second braking burn. The orbital period decreased to 84 minutes. Although the burn decreased their speed at the new apoapsis burn point, it increased their speed at periapsis 180 degrees downrange. This gave the crew a better view of the terrain at periapsis, although their speed was somewhat faster.

"Our orbital parameters indicate this planet's gravity and therefore its mass is slightly less than Earth's, but not by much," said Kevin. "See if we can pick out a flat landing site. We have plenty of time, so we'll stay in orbit until we choose our new home. Our orbit is now an ellipse, so we will be lower at the periapsis to get a closer view."

"The planet looks something like Earth," said Jason, "with clouds across both the land and sea masses."

10. Dreamland

"I think this planet is rotating under our orbit," said Michael. "That is one of the four conditions for a magnetosphere. We must be inclined to its plane of rotation because the land mass below us is different from our last pass."

"Great," said Kevin. "We can eventually see almost all of the planet to find the best place to land. We are not in any hurry.

"Also, it looks like it is not tidally locked to Proxima Centauri like Proxima b is because its rate of rotation is much faster than its orbital period."

"Half of this planet is ocean," said Jennifer. "Hard to tell for sure from up here, but the land looks mostly flat and green."

"You are right," said Samantha, excitedly. "Maybe this land is covered with plant life."

"It clearly has clouds!" shouted Jason. "That must mean it has water vapor in its atmosphere. It is unbelievable that we should be so lucky to find a planet so much like Earth. This is a one in a hundred shot."

"That means it must have some kind of life form," said Kevin. "But maybe not like Earth's."

"Well, gang, shall we pick out an island or a continent?" asked Kevin, not so calmly.

"Let's land on a continent," said Candice. "Then we can explore without needing a boat."

"Great idea," echoed Jason. "But we have so much to choose from."

"The planet seems to have ice caps, so we will land in a temperate zone near the equator and far from the poles," said Kevin. "We don't know its temperature, but the temperate zone gives us our best chance."

"But Kevin, don't we have to worry about our Thermal Protection System's vulnerability due to the patch?" asked Jason.

"Yes," said Kevin. "I've given that a lot of thought and think I have a solution. If we increase our altitude enough, we can kill off most of our speed on the way down, so that our entry velocity is low enough to minimize atmospheric drag."

"Skipper, that is brilliant! It should work!"

"Orbiter, on the next orbital pass, synchronize a burn to kick our apoapsis high enough to restart the Raptors, first to kill off enough horizontal velocity, then the final braking maneuver to reduce our vertical descent rate to less than 1,000 miles per hour as we enter the atmosphere. Time the burn such that we land near the coastline of the upcoming land mass north of the equator. Aim for the grassland we saw on the last orbital pass. As we encounter the atmosphere, throttle down the engines to counter the planet's gravitational pull for a soft vertical landing."

"This is so exciting," declared Candice.

"Crew," commanded Kevin, "put on your spacesuits and secure yourselves into your Dragon couches for descent and landing. I'll restore the covering over the cupola for entry."

The deceleration again caused a ½ G force on their bodies during the three-minute burn to raise their apoapsis., The engines then shut down as they coasted higher before restarting near apoapsis. They experienced 3 Gs deceleration force as they felt the braking maneuver. At atmospheric entry they felt slightly more than 1 G. The modified duct tape held easily. Starship successfully touched down safely.

"Starship has landed," announced Kevin, mimicking Armstrong's first words on the Moon. They all let out a loud cheer of relief and gratitude. Michael then led them in a prayer of thanks to God for a safe landing. He administered the holy communion he had brought, thanking God for safe passage, just as Buzz Aldrin did when Apollo 11 landed on the Moon.

Kevin demanded that all remain in their couches until he ascertained their landing craft was stable. Then he opened the hatch at deck 12 and carefully emerged onto the ladder rungs on the side of Starship. He gasped at the beauty of the landscape of their new home as he descended to the surface. "This is REALLY one giant leap for mankind."

Kevin carefully carried a small American flag to the surface and, with a broad smile, easily planted it into the regolith alongside the Spaceship.

The others could not wait to follow as they descended single file. Mike helped Sam gingerly negotiate the ladder rungs, as she was now almost eight months pregnant. As they descended, they were blown away by the scenery. They were euphoric.

"Awesome. This looks just like Earth," exclaimed Samantha into her spacesuit communication system.

"This is beautiful," added Jennifer. "It looks like a savannah on Earth. And my suit thermometer indicates a temperature of a balmy 79 degrees."

"Don't forget its orbital plane is more inclined to the Alpha Proxima ecliptic than Earth's plane is to the Sun," said Kevin. "So, it likely has seasons more pronounced than Earth's. Therefore, its summers may be warmer than Earth's and its winters colder. I think we landed during late spring."

"It takes my breath away," said Candice. "Look at the beach over there."

"And check out the hills in the distance," said Jennifer.

"And the blue sky with white and grey clouds," said Candice.

Samantha could not contain herself: she began singing *What a Wonderful World.* Eventually they all joined in. A bit corny, but they didn't care. They all loved corny.

"We have three Suns!" yelled Jason. "One big one and two small ones on the horizon. I can see all three in the daylight. The small ones are tiny, yet very bright."

"I think the gravity is a little less than Earth," suggested Mike as he began stumbling on the surface, his legs not used to the nearly 1 G force.

"The planet must have an abundance of water to grow this lush grassland," said Kevin. "We may have struck it rich."

"Do you think the atmosphere has enough oxygen?" asked Jennifer.

"Hard to say," said Kevin. "But it must contain carbon dioxide for this grassland to look so healthy. Make sure your spacesuits are secure until we find out for sure.

"It looks like we have several hours of daylight to set up our paraphernalia," observed Kevin. "Let's climb back up and retrieve only the equipment we need immediately: the 3D printers, the Geiger counters, the brick-making additive, the solar panels, Sam's plants, and Mike's tools. We'll bring the rest down after we build our habitat."

"Let's include the robot," suggested Jason. "We can put him to work for us."

"You're right, Jason," said Kevin. "I forgot about that asset."

Kevin, Mike, Candice, Jen, and Jason climbed up the rungs back into Starship. Candice harvested the mature fruit and vegetables and retrieved the roots of two of Sam's precious vegetable plants: Romaine lettuce and Chili peppers, from their hydroponic housing and carefully carried them down to the surface. Sam had already selected a place for their garden. Mike brought a spade and a trowel. Kevin and Jason each carried one of the 3D printers.

Jason went back to retrieve the robot and struggled to bring him down the rungs of the ladder, carrying the five-foot eight-inch 128-pound humanoid over his shoulder. "This guy is heavy," he grunted. He did not realize that he could have programmed the bot to scale the down ladder automatically. He was impressed how much it looked like a real human. It was 5-feet 8-inches tall with arms and legs, fingers with opposable thumbs, and as much flexibility as a human. Its face was pleasant looking but did not show different facial expressions.

Inside its head was a wonderful high-speed computer which was susceptible to all artificial intelligence programming. It responded to verbal commands through its receivers in place of ears. It received visual images from sensors instead of eyes routed into its AI computer.

Jennifer carried sacks of the brick-making additive. Kevin went back for the solar panels.

"We may not need these solar panels at first," he said as he deployed them facing their new Sun. "This is just in case we need additional power beyond that supplied by the nuclear engine. But eventually we will have to replenish and supplement the energy from the nuclear engine."

Jennifer joined Candice in retrieving a tomato plant, red Russian kale, and strawberries from the hydroponic container.

Meanwhile, Mike used a 3D printer to duplicate three more trowels from the original one. Then he began digging into the grassland with the spade to produce the 25-foot square dirt patch Sam had chosen for their garden. Sam, Mike, Candice, and Jen began using their new trowels to prepare the soil for Sam's plants.

"This soil is great," said Sam. "Looks and feels a lot like soil back in Iowa."

"We'll have to bring water down from the hydroponic garden as soon as we have these in the ground," said Sam. "Let's hope that's all they need for now. We could get more water from the Starship, but we will have to find a water source soon."

"Mike, you and Jason use the 3D printers and additive stuff to see if you can make bricks for the habitat," said Kevin. "The brick template has been preprogrammed in each printer."

Kevin deployed the two solar panels to catch Centauri Proxima's rays. He hooked them up with the sidereal drive to follow Zeus's rotation. "Hey," said

Mike. "we don't know Zeus's rate of rotation to program this drive."

"Just set it up for 24 hours for now until we can measure it over the next few days," said Kevin.

"Zeus is a poor name for this beautiful planet," said Jen. "Let's think of a better name."

"How about Paradise?" suggested Candice.

"Or Heaven?" said Jason.

"Or Shangri-La," offered Sam.

"I like Dreamland," said Mike.

"Good one!" responded Kevin.

"Okay. This planet is Dreamland, unless we come up with a better name," said Jen.

"Now, let's see if we can use this soil to construct bricks for our habitat before the Sunsets," said Kevin. "This habitat must protect us from the solar radiation in case Dreamland does not have magnetic fields and enough atmosphere to defect and absorb the radiation before it reaches us."

"While we're at it let's clear another patch for our heat sink and cold sink to regulate the internal temperature of the habitat," suggested Mike.

When they finished the garden, Mike created another soil patch half as large as the garden.

"Will this suffice for now, Kevin?"

"Perfect. Once we rig up the solar panels, we can evaluate it for energy collection, storage, and release."

Jason volunteered, "Let me see if I can program the robot to help construct the bricks.

"Hey, I think he's been preprogrammed for this task. He seems to know what to do. He only needs a little guidance. Let's create a pile of soil next to the binding hydrogen polymer additive and the robot can manage it from there using the 3D printer. Check him out. He is amazing. He knows what to do. Do we have a name for him?"

"Elon Musk called him Tesla Bot or Optimus," said Kevin.

"He deserves a better name," said Jason. "How about Homer?"

"That works for me. He'll be your pal, Homer."

They all pitched in gathering soil for Homer to feed the printers, adding the binding polymer to the mix.

"Seems to be working perfectly for creating substantial bricks," said Mike.

"I think it will take Homer only a couple of days to have enough bricks to build our igloo," said Kevin.

"We can let him work all night," said Jason. "I'll bet he won't need rest or intense supervision."

"Jason, you monitor Homer while we help Sam finish her garden."

"This soil looks great," said Sam. "Looks like the stuff on my farm back home. Hope my plants like it."

Candice climbed up the ladder and retrieved four containers of water from Sam's hydroponic system on board. She carefully applied water to each plant.

"I hope they like their new environment as much as I do," she said. "We will know by the end of day tomorrow whether this works. If not, we will have to forage to see what this planet may provide. Anyone hungry? If we don't continue the hydroponics, we have only enough produce in the Spaceship to last a couple of weeks."

"Wait a minute," said Candice. "I think it's beginning to RAIN!"

"I noticed the clouds forming," said Jason. "Could we be so lucky?"

"Let's catch some of it with this cup I brought from our galley, take it back inside, and see if it is really water," said Kevin.

Candice did exactly that. She went inside Starship, removed her helmet, and sipped the rain liquid from the cup. "Tastes like real water to me. Hope I don't get sick." She yelled to Kevin, "Kev, it IS water. I'm sure."

"That's fantastic!" he responded.

They looked at the sky to watch the beautiful rain fall. Rays of Proxima Centauri began peeking through the clouds. "A RAINBOW!" yelled Jen.

Sam wanted to dance as she again remembered the lyrics to *What a Wonderful World,* so appropriate to her beloved Dreamland.

"Okay, now we need to create a cistern to catch the rainwater," said practical Kevin.

"I'll make a crude one from the tarpaulin covering the rover," said Mike.

"Candice, could you construct the cistern?" asked Kevin.

"Sure. Just retrieve the tarp for me."

"Okay gang," said Kevin. "It looks like the Sun is setting. Let's climb back up and get out of these spacesuits. After a good night's sleep, we can tackle beginning to build our habitat early tomorrow. Nice to have Homer working through the night for us."

"Kevin, how will we know whether we can breathe the air and eliminate the spacesuits?" asked Jennifer.

"Well, we can experiment somewhat, but even if the air is breathable, we have to keep the suits on until we learn about the radiation."

"How do we do that?"

"We have two Geiger counters stowed in Deck 8. Tomorrow, we will deploy them to measure how much harmful radiation we will encounter. After a couple of days, we will learn our degree of risk and whether Dreamland has enough atmosphere and magnetosphere to protect us."

Before they went to their pods, Samantha grabbed her guitar, turned on her laptop and led them all in a series of songs of joy. They began with their theme song,

Fly Me to the Moon, and concluded with Michael's favorite hymn, *How Great Thou Art*.

That night they all had trouble sleeping. Their excitement knew no bounds. Candice stayed up late gleefully working on her journal.

The next day, the crew scampered back down to the surface. They discovered Homer diligently producing the bricks in abundance. They enjoyed using these to assemble their new home brick by brick. They completed the habitat construction in several hours. They all used the spade and trowels to cover the igloo with regolith dirt to protect against radiation.

Kevin examined the solar panels and sidereal drive. He timed the rotation rate of Dreamland to an estimated 1,340 minutes.

"Now let's see if this oxygen generator works as well with this planet's regolith as it did on the Moon. I'll hook it up to the solar panels' power."

Mike adjusted the solar panels and sidereal drive, updated to the 22.3-hour day to directly capture the Sun's rays and ran the wires into the habitat.

"It seems to be working," said Mike.

They all entered their new igloo and found comfort in sitting down.

"I'll try breathing the generator's oxygen," said Kevin, as he gingerly removed his helmet.

"Yes, it is definitely oxygen. That means the soil picked up oxygen molecules. So, the atmosphere does contain some oxygen, but we don't know how much. It would have to be between 17.5% and 23.5% with the rest an inert gas like nitrogen for us to breathe it without additives.

"I'm tempted to go outside, remove my helmet, and try it."

"Isn't that dangerous?" asked Jen.

"Not really. The supply of oxygen in my lungs now will suffice for a minute or two. The only risk would be if the other components of the atmosphere were toxic. And that is unlikely, based on the flourishing condition of the grasses."

Kevin had talked himself into the experiment. He exited the tunnel airlock, carefully unfastened his helmet, and took a short breath. Then, a deeper breath. He was fine. He quickly reentered the tunnel airlock, beaming.

"The atmosphere is breathable! It is a miracle. It's like we brought Earth's atmosphere with us."

"What a discovery this planet is!" exclaimed Jason.

"And the Geiger counters indicated no more radiation than on Earth."

"I should send NASA a communique about where we landed and how fantastic this planet is," said Kevin. "As far as they know, we tried to land on Alpha Proxima b."

"Let's not rush it," suggested Mike. "We have plenty of time since they won't get any message from us for over four years anyway."

"Sure, we can wait."

"Mike, can you fix the rover, so we can explore with it?" asked Candice.

"I think so."

"But Kevin, can't we explore on foot for a while?" asked Jennifer.

"Yes, but we need to stay in our spacesuits for now, both inside the hab and outside, until we are sure we don't need them for total life support, not only oxygen and water, but radiation protection, constant air pressure, and unanticipated temperature extremes. We don't yet know how cold it could get after Sunset. If the atmosphere is dense enough, the temperature could moderate much like Earth. But we'll learn that during our first full planet day.

"Mike, do you think you will need help to fix the rover?"

"I doubt it. It looks like it is just bent from the collision. The motor was not damaged, and the transmission looks okay. After I unfold it, I will let you know if I need help."

"Okay, crew are we ready for a short excursion this morning?" asked Kevin.

They replied in unison, "Yes!" as if they were kids replying to an invitation to a trip to Disneyland.

"Okay, make sure your spacesuits are ready."

One by one, they exited the igloo and waited for their leader.

"Let's go to the beach," suggested Candice enthusiastically.

"The beach it is," responded Kevin.

"I can't wait to wash the clothes I've worn for the past eight years," said Jen. "Hope the ocean water can handle it, and they don't fall apart."

"We may all be wearing loin cloths," added Candice.

"Works for me," smiled Jason.

"Mike, can you hear us?" asked Kevin as they began their first exploration.

"Loud and clear," responded Michael as he began his work.

"Let us know if our conversation begins to fade."

The crew was enthralled as they approached the beach.

"Sand like on Earth," exclaimed Candice.

"And these look a bit like seashells," added Jason. "Wonder what kind of sea creatures live here."

"I wonder how long they've been here and what their evolutionary history is," added Kevin.

"The water, if that is what it really is, looks just like the ocean on Earth," said Samantha. She scooped up a handful, opened her helmet, and tasted it. "I think it IS seawater, salty like Earth."

"Anyone for fishing?" asked Kevin.

"We need to build a boat and get some bait," suggested Jen. "I could go for some fresh seafood after more than eight years of veggies and prepackaged dried packets."

"We don't have hooks, lines, or fishing poles," said Jason. "Besides, we don't even know if there are fish in this ocean."

"It's pretty likely there is some kind of sea life, based on those things that look like shells," responded Kevin. "And we may not need hooks, lines, and poles. We can fish like natives. All we need is to weave a net, throw it in, and see what we drag up."

"Let's go back to camp and see if we can weave a net," said Jen.

"We need some swimsuits," said Sam.

"Let's go skinny dipping, Jay," said Candice, almost giddy.

"Too bad we've been ordered to stay in our spacesuits," answered Jason, smiling.

"Bet we could float in our spacesuits, especially in this salty water," added Jen.

"Don't try it," commanded Kevin.

"Lighten up, Kev," admonished Jennifer.

"Commander, your signal is weakening," announced Mike.

"How are you doing?"

"Great, I've finished, and the rover is humming."

"Okay, kids. Enough frolicking for one day. We are heading back."

Back in the habitat, they retrieved the rest of their paraphernalia from Starship: clothes and sewing equipment, iPads and computers, medical equipment, sleeping pods., and the table and chairs. But they decided to leave the table, chairs and Peloton outside due to the cramped space in the igloo. Jennifer found enough twine and began weaving a small fishing net.

"We need more room and privacy," said Sam. "Kevin. Let's build the other two igloos. Do we have enough additives for it?"

"We sure do. Jason, do you think Homer is up to the task?"

"Of course. He likes to keep busy. I'll get him back to the brickmaking tonight. But we need more regolith soil for him."

Mike and Jason began to dig another pile of soil.

"Based on the Geiger counter data and what we learned about the atmosphere, I think we could venture out without our spacesuits for a short time tomorrow," said Kevin.

"We need to find potable water."

"There has to be a river between those hills and the ocean," said Jen. "Let's go back to the beach and look for a river basin."

"Good idea," concurred Kevin. "I have a couple of containers if we do find water."

"Since you guys don't want to go skinny dipping, I can make swimsuits and at least get our toes in the ocean," added Candice.

"I'll help," said Jen.

The next morning, Kevin, Jennifer, and Candice donned their new swimsuits under their clothes and ventured to the ocean shore. Samantha looked forward to doing the same after her baby was born. Mike and Jason began building the second habitat with Homer's new bricks.

When they reached the sand, Candy and Jen took off their outer clothes exposing their new swimsuits. They left their shoes and their other clothes on the sand. Kevin couldn't help admiring their trim, firm bodies. The daily exercise obviously was paying welcome dividends.

They began wading through the water, smiling, splashing, and giggling.

"Come on, ladies," said Kevin. "We have to find a river. Let's start moving north along the shore in the direction of the hills."

"Okay, spoilsport, but this water is great," said Jen. "We'll get back in when we return."

Many minutes later, Samantha remarked, "I think ahead is the mouth of a small river."

"You're right," said Kevin. "Let's follow it upstream a bit."

"We don't have shoes," said Jen. "I'm not walking through that tall grass barefoot."

"Okay, I'll do it alone."

He followed the river stream and filled his containers. After tasting the water, he brought the containers back and handed them to Jen and Sam. Each took a long drink, then let Candice do the same. Kevin refilled them and returned to the shore.

"Okay, let's go back to camp. We've been in the Sun long enough."

"We're enjoying the water," said Jen. "We'll go back the way we came.

"Come on Kev, take off your shirt and enjoy the Sun. Show us your beautiful physique."

Kevin proudly displayed his six-pack.

"What is that stuff growing on the back of your neck?" asked Sam.

"just a little dry skin."

"Let me see," demanded Jen. "I think these are basal cells. Did you use Sunscreen playing golf all those years?"

"No. I have dark skin. I didn't need it."

"Everyone needs to protect against the Sun, especially when you play golf for hours at a time. When we get back, I'm going to check it out."

They found their clothes and shoes and headed back. They arrived at the habitat as Mike and Jason were busy building the second igloo.

"We found a small freshwater river just 1,200 yards north of here," announced Kevin. "Have a drink."

"This is great," said Mike. "I can build an aqueduct with a tube and a pump manufactured with the 3D printer. So, we no longer have to depend on rainwater."

"We'll have running water inside the igloos?" asked Sam.

"Once I build a plumbing system," answered Mike.

"With real toilets?"

"Yup. If I am successful."

"Okay, Kev. Let's have a look at that neck," said Jen. "I'll do a biopsy."

"Ouch!"

"Sorry. Guess I should have started with an anesthetic."

Jen took the sample to the microscope. Then returned.

"Just as I thought--basal cells."

"I'll do a Mohs procedure. You'll be fine."

"A what?"

"A Mohs. I cut out a portion. Assess it. If there are no more cells around the edges, I sew it up. Otherwise, I keep cutting."

"Sounds like fun, NOT."

"If we don't arrest it now, it could turn into serious cancer."

"Okay, you're the doctor."

"You could have avoided all this with a little Sunscreen each time you played golf."

After the anesthetic and two iterative cuts, Jen declared all clear and stitched the region.

“Keep the region bathed in this Vaseline. In two weeks, I’ll remove the stitches. But no more Sun without protection, young man.”

“Yes, Mother.”

When they completed the second igloo, Mike tested the collection cistern. The beautiful rain was coming down again, but much harder now. Candice stayed outside to enjoy beating on her face.

The next day, Sam wanted to explore again, this time searching for other plants besides the grasses. “Can we take the rover?” she asked Kevin. The others stayed to help Mike and Jason construct the third igloo.

“Sure. Hop in. We’ll head back to the river.” When they reached the river, they turned west to follow the bank.

“This looks a lot like Iowa, except the grass is much taller,” said Sam. They saw some leafy plants and bushes on the banks.

“Kevin, look at that bush. I think those are elderberries. Let’s pick them and see.

“Mmmmm, they are good and sweet. Try one.”

“Thanks Sam. These are wonderful.”

“Let’s take some back for the colony to try.

"Kev, do you think there are any animals in the river?"

“It is possible. It depends on how long our planet, er, I mean Dreamland has been evolving. Remember, it took Earth billions of years to evolve plants, then fish, then animals of every kind.”

“So, you don’t believe in the Bible that says God created the Earth and all in it in just seven days?”

“No, but I do believe that God created the Earth and all our life forms, but He did it in eons, not days. Plants were created first, perhaps more than a billion years ago, three billion years after Earth was created. Earth got its oxygen from the plants. Primitive forms of sea animals followed, but the first animals that walked the Earth are believed to have begun 360 million years ago. So maybe Dreamland is undergoing a similar evolution. We just don't know what stage it is in. We don't even know how old Dreamland is. By locating various plants today, we may be able to surmise something of Dreamland's evolution status. So far, all we've seen is grass and a few berries. But those things that looked a bit like seashells on the beach may be a better clue."

"So, you do think there are animals on Dreamland?"

"If the things we found on the beach are seashells that housed sea life, then, absolutely. But that doesn't mean Dreamland possesses sophisticated land animals."

"Kevin, look! There is a tree."

"That indicates some degree of evolution, but that is not surprising."

"I think I just saw something slither into the water."

"Now that would be something. The first salamanders appeared on Earth about one hundred million years ago. So, if that was an amphibian, Dreamland may be more geologically advanced than we think."

"Aren't those lily pads in that alcove of the river?"

"Yes, I think you are right. That is consistent with the existence of frogs: maybe one hundred million years of existence.

"I think we should be heading back. But we learned something about Dreamland to share with our crewmates."

"Thanks, Kev. This was invigorating."

When they returned, Kevin hooked the rover up to the solar panels for recharging. Sam bubbled with tales of their adventure and findings. Then she checked on her garden. She was overjoyed.

"Our plants are flourishing; they must love this soil. We are not going to starve. The lettuce looks green and heathy. And the Chile peppers are growing. Let's bring the rest of the plants down and transplant them."

Mike and Jennifer retrieved the radishes, Chinese cabbage, Russian kale, tomatoes, Walden's green lettuce, Mizuna mustard, and strawberry plants and added them to the garden. Samantha beamed with pride. They all dined on the vegetables they had retrieved from the hydroponic garden, now knowing they no longer had to ration their food. They had a celebratory banquet.

Jason said, "Tomorrow I intend to explore with our rover. Anyone want to come along?"

Candice said, "Absolutely."

Mike needed to oversee Homer's completion of the third habitat and begin construction of the pipeline.

Jennifer said, "I will stay here with Sam, just in case she needs me for any prenatal care. It is time for another ultrasound."

"Jason, where do you want to investigate?" said Candice.

"Let's head toward those hills. I think the river begins there before it wanders back southwest toward the ocean. Kevin, what range does the rover have with a full charge?"

"Today we traveled twenty-two miles and used only 40%. So don't go farther than twenty-five miles and you should be okay. But if the terrain gets steep, just monitor the battery gauge to make sure you have plenty of margin to return.

"I wish we had my Tesla," said Kevin. "It travels 320 miles on a single charge."

The next morning Jason and Candice excitedly began their first exploratory trip on Dreamland. They took along their spacesuits just in case. They headed northwest in hopes of not having to forage the river. The rover easily reached a speed of eight miles per hour for the first thirteen miles before reaching the foothills. The grade was a gentle 3-4 degrees at first. Then they encountered a stand of small trees, which became denser as they ascended. They stopped with more than 60% charge left and carefully continued on foot for several hundred yards. They were euphoric to be in a tiny forest, admiring the canopy overhead and the leafy floor at their feet. Suddenly Jason held up his hand to stop. He had noticed the leaves in motion ahead. He whispered, "Candy, I think I spotted a small animal."

Yes," Candice whispered. "I see it. What is it?"

"I've never seen it before. It looks like a squirrel, except it has a smooth shiny green skin and tail instead of fur and has a flat nose."

"Should we try to catch it?"

"No. I don't like the looks of those teeth. Besides, we have nothing to keep it in. Let's just follow it from a distance."

But the creature heard them and very slowly moved away as if it didn't care about their presence. They stopped following it.

Their subsequent walk was uneventful, so they returned to the rover with a wonderful tale for their crewmates. They were giddy with the news when they reached the colony. Both tried to talk at once even before recharging the rover.

"Gather 'round guys. We have an important discovery to share," said Jason.

"Dreamland has evolved much more than we thought," blurted Candice. "It has land animals."

"Are you sure?" asked Kevin. "What did you see?"

Both excitedly described their findings.

"That means Dreamland must have been formed billions of years ago, which we already knew from this environment. But there's no telling how far it has come in its evolutionary maturity. It might be as mature as Earth. Possibly even more."

"Might there be predators?" asked Jen.

"What about dinosaurs?" asked Samantha.

"Are we in danger?" asked Jennifer. "We don't have any weapons."

"None of us can answer those questions," said Kevin. "At least not yet."

"Could there be intelligent life?" asked Jason.

"It is possible," responded Kevin. "We may soon know."

"Wow," said Sam. "This is all so scary and crazy."

"And exciting," added Jen.

"So, we just sit here and wait to see if something or someone materializes?" asked Candice.

"Unless we continue to explore," said Kevin.

That night all six slept uneasily.

The next day they moved into their three completed igloos. Kevin and Jen housed the table and chairs, anticipating using them for colony meetings and playing games. Mike and Sam took the one adjacent to the garden and housed the Peloton. Jason and Candice's igloo housed Homer.

Mike and Jason couldn't wait to explore more. But Mike agreed to work on the aqueduct so necessary for their water supply. The skies were clear and blue. They couldn't count on rainwater filling their cistern. And Jason had to program Homer to create the piping. Candice and Samantha were happy to stay in the colony to tend to the garden.

But first, Candice wanted to try her fishing skill. "My fishing net is complete," said Candice. "I want to try it out." She and Samantha brought it to the ocean, tossed it in and waited.

"I think there is something in the net," Candice exclaimed. They dragged the net to shore and examined the 14 -inch creature. "What is this thing? It looks like a small grouper, but it has no scales. It has gills and fins, so it must be a fish."

"Now what do we do with it?" asked Sam. "I'm sure not going to try to eat it."

"Let's wrap it up in the net and take it back to the colony. We can preserve it in the cistern until the others return. They may know what to do."

Meanwhile, Kevin agreed to head another expedition, but this time he took Jennifer along and decided to head south. They found nothing but grassland for twenty-five miles, so he turned around back to the colony to recharge the rover. After recharging, he and Jennifer retraced Candice's and Jason's route northwest to find the edge of the small forest. There they left the rover and ventured into the trees.

"Kevin, I see some leaves moving just ahead," said Jen.

"Let's check it out," said Kevin. "I think it might be a snake."

"Not too cool," replied Jen.

"It looks like snakes on Earth. But I don't recognize it as a North American variety. It is small and blue green. But it may still be poisonous. So, we'll avoid it."

Kevin searched the limbs of the trees overhead to ensure no snakes were hanging around.

They continued walking through the trees until they came upon a small pond. It was apparently fed from an underground spring. Several creatures were drinking from the water.

"Kevin, what are these things?"

"I don't know, but they look docile enough."

Gathered around the pond were three varieties of small animals. None appeared frightened by their presence. One group of four looked like small cats with tiny ears and eyes but had no fur. Another group looked like pygmy possums, but again had no fur. The third looked like tiny monkeys, maybe marmosets, with flat faces and no fur. All three groups had greenish skin. They ignored each other and showed no fear, letting Kevin and Jen approach. They all continued to drink the water.

"Kev, they are so cute, and they are not afraid of us. Do you think we could adopt them as pets? I would like to take them back to the colony and show the others."

"No, Jen. We don't know anything about these creatures. We will leave them alone."

"Why don't they have any hair?"

"Perhaps their evolution didn't require it. Either it doesn't get too cold in this part of Dreamland, or their body temperature suffices. Don't forget that as humans evolved, most of us lost almost all our body hair."

"And why are they green?"

"I wish I knew."

"So, do you think their evolution went beyond that on Earth?"

"Possibly. If so, I would like to see more species here."

"Kevin, this is getting a little creepy. Let's head back."

Kevin reluctantly agreed. They got back into the rover and drove to the colony. But Jen couldn't wait to tell the others of their discoveries.

Before they began their tale, Candice and Sam displayed their prize fish in the cistern.

"Looks like a strange fish to me," said Kevin. "No scales. But I don't think we should try to eat it until we learn more about it. Please take it back to the ocean. It could die in fresh water. And it's probably polluting our cistern."

"Candy, you should have seen the animals we discovered at the river," said Jen. They looked like marmosets, small possums, and domestic cats, all drinking from the river. But they had no fur and had greenish skin. They were not afraid of us. But ***I*** was afraid of ***them.***"

"Now I'm even more worried about predators and dinosaurs," said Sam. "How can we defend ourselves?"

"Let's relax," said Kevin. "We have no reason to suspect we could be under attack. The creatures on this planet could all be self-sufficient and not be motivated to harm us."

"Famous last words," quipped Candice.

"Sounds like we entered the twilight zone," said Jason.

"But great stuff for my journal," said Candice.

"How is the aqueduct coming?" asked Kevin.

"Homer is fantastic," responded Jason. "He has half the piping finished."

"We could begin laying it tomorrow," added Michael. "I'll start working on the design of the pump. Then the internal plumbing for the igloos."

"Nice work, guys," said Kevin.

"And how's our garden coming, ladies?"

"The roots have held beautifully, and the vegetables and fruit have started to grow," said Samantha. "I think we can begin to harvest in two months."

"Wonderful! We have a real surviving colony."

"Sam, how's the baby doing?" asked Candice.

"It's beginning to kick more strongly. I think it's quite healthy."

Michael smiled proudly. "An ideal pregnancy."

"The ultrasound looks great," said Jennifer.

"When are you guys going to get started?" asked Sam. Our bambino will need some companionship."

"We'll get there soon enough," said Jen. "Be patient." Jason and Candice nodded their agreement.

The next day as Mike, Kevin, and Jason began laying pipe to the river, Homer was as busy as ever finishing the job of creating the pipe segments.

Kevin wanted to continue to search for more life in the forest but was busy with the aqueduct. Sam decided not to venture out. Jen and Candy stood by her.

11. Visitors

"Kevin," yelled Jennifer. "Come outside! Look at the sky!"

Overhead he saw the same shiny disk they observed in their Spaceship cupola two years earlier. It first hovered overheard; then began to descend near their encampment.

"Wow. We have a visitor. It looks like it's going to land. Jen, quietly alert the rest of the colony."

They all came out aghast at what they were seeing. Candice and Samantha were especially frightened and began to cry. The men froze in their steps as they looked on in wonderment.

"Calm down, everyone," said Kevin. "We are about to experience a close encounter of the third kind. Do not cower. Let's meet these creatures with dignity and hospitality."

"Are you crazy?" asked Sam. "They are here to kill us or eat us. We have no weapons to defend ourselves."

"Right. We don't," said Kevin. "So, let's assume they are friendly and want to meet us in peace."

They all watched as the craft very deliberately descended, touching down not more than one hundred feet in front of them. Well, it didn't actually "touch down," but hovered four inches off the ground. Yet it made no noise. A hatch slowly opened, and six aliens climbed out and stood before them. They are all approximately five feet tall and look a bit like humans. They each have two arms, hands and fingers. They stand tall on their two legs. Their heads are a bit large for their bodies by human standards. Their heads have no hair. They have shining black eyes and human-like ears, mouths, and noses. Their skin is greenish, but their bodies are covered with blue jumpsuits. A small tube extends from their jumpsuits into their nostrils. Each is wearing boots on their feet. Although their countenances are similar to each other, it is not difficult to tell them apart. The three females are obvious. Their jumpsuits are form-fitting, displaying their attractive curves. Although they apparently are wearing no makeup, their facial features are beautiful with small lips and noses, high cheekbones, and winning smiles. The tubes in their nostrils do not detract from their beauty. Their thin eyebrows frame their best feature: the prettiest eyelashes the colonists had ever seen. And they know how to flirt with them. All six are smiling, revealing perfect, white teeth. They appear to be not older than thirty.

One began to speak. "Do not be afraid. We come in peace." Kevin assumed this was their leader.

"You know our language? How is that possible?" asked Kevin.

"We learned not only your language but your mission objective and your individual demeanors," answered the alien. "You may remember that we rendezvoused with your spacecraft two years ago. We spent a great deal of time examining your minds, bodies, and motivations. It was then that we concluded to let you fulfill your objective and determine whether you were able to find a suitable planet for your colony. You have chosen well.

"We already know your names, so let us introduce ourselves. You may call me Orka. This is Ampa, Solpe, Perna, Nambe, and Twinto, They are not our real names of course, but anglicized for your benefit. You would not be able to pronounce our real names."

"Thank you, Orka," said Kevin. "You have put us at ease," he lied. "Where are you from?"

"Our home planet is in a solar system your astronomers call TRAPPIST-1."

"May we ask you questions?"

"Of course. That is why we landed – to help you establish your colony. But first you need to know about us and learn to trust us. We can tell that you fear us and question our motives.

"First, I will tell you a little about myself. I am the captain of our spacecraft, the elected leader of our expedition, and the being responsible for everything we do. Kevin, we have similar roles. I have been trained for these roles on our home planet. We have an assigned mission to investigate any visitors to our vicinity of the cosmos and ensure no such visitors have any objectives to invade, conquer, or invoke any hostility in this proximity of our galaxy. You humans would call us the Space Patrol. That being said, be assured we welcome you in peace. Now, more about us individually.

"Let's begin with my mate, Perna."

"Welcome to the Alpha Centauri system. I believe you will enjoy living here. In many ways the environment on this planet is much like your Earth's. Orka and I have been mates since puberty. On our planet, we mate for life. We have several children whom we raised during the first formative segments of their lives. Each is now raising families of their own. We are all self-sufficient and self-sustaining. But we continuously cultivate and enhance our relationships with each other as well as our mates. Although we five have accepted Orka as our leader, we are all equal. Each of us has our own role with specific duties to benefit the mission and each other. My role is to treat any of our medical needs. I obtained my medical training on our home planet which human astronomers call TRAPPIST-1e. Jennifer, it would seem you and I have much in common, professionally. We should spend some time together comparing the design of our anatomies, our medical procedures. and our pharmaceuticals."

Jennifer was listening intently and began to feel more comfortable. She was anxious to meet with Perna personally.

"Ampa," said Orka. "Tell our Earth friends what you do."

"I am the official engineer for our spacecraft and navigator for our travels. I too am from TRAPPIST-1e.

Our planet is similar to Earth but is almost forty light years from your Sun. It has oceans and is a rocky planet slightly smaller than Earth. So, TRAPPIST-1e has some similarities to your newly chosen planet, which I learned you now call Dreamland. One significant difference is the evolution of life. TRAPPIST-1e and our life forms have been evolving for more than four billion of your Earth years, much longer than Earth's life forms. Our planet was born almost eight billion years ago. As a result, our civilizations are considerably more advanced. Another difference is the atmosphere. We don't breathe oxygen, but rather methane.

"We have been traveling throughout several stars in this galaxy which you call the Milky Way. Here we are learning about the many habitable planets which might harbor life forms. Like humans, we know that water is a necessary element to sustain life of all kinds. But you would be surprised as to the number of planets in the habitable zone of their stars whose evolutions are far from being mature enough to sustain life as we know it. You are very fortunate to find this planet in the proper stages of its evolution. We can tell you that very few such planets are at this stage in their evolutionary juncture. In our TRAPPIST system three of our seven planets sustain life. In your Solar System, of the eight planets, only Earth has this attribute."

Both Mike and Jason were bursting with a million questions.

Nambe was next to address the colonists. "Ampa is my lifelong mate. But we did not meet on TRAPPIST-1e. Instead, we met on TRAPPIST-1f, just after our spacecraft landed. After a whirlwind romance, we married, then had five children. That was more than three centuries ago, using Earth measures of time.

"I, like your own Samantha, am an agronomist. But the varieties of our edible plants differ from the ones you are growing here. They are all a staple of our diet. We also grow them aboard our spaceship and transplant them on several of our destination planets. You might say my role is to be the breadbasket of our team. We learned long ago the benefits of veganism. We have no need for beef, poultry, or fish. Once we realized this, we were able to reduce our energy needs, reduce toxic waste, and minimize carbon dioxide and nitrous oxide air pollution. This also hampers bacteria mutation, minimizes medical problems by reducing disease risk, improving our immune system, and helps to preserve our water supply. Veganism is a win-win for us and our environment."

Solpe took the next lead. "Welcome, visitors from Earth. After observing the conflicts on your home planet, I never thought I would be uttering those words. I hope we can enable you to be comfortable with our presence. Our surveillance of you personally has provided us with acceptance of your team and its goals.

"My role is multifaceted. My education includes various technologies. You call them artificial intelligence, computer science, and engineering. Our society actually learns the basics of these disciplines in elementary education. I don't mean to be disrespectful, but our advanced learning builds on these disciplines to go well beyond these basics. So, in a sense, you might call me a jack-of-all-trades. Maybe a conglomerate of Mike's and Jason's' professions combined into an advanced state."

Twinto took over. "Solpe is my loving mate. Like Ampa and Numbe, we have been together forever," smiling and winking at Solpe. "Or it seems like forever."

"Like Solpe, I have multiple roles: housekeeping and exercise training. I organize and maintain our spacecraft with an iron fist. No one moves any of our equipment, foodstuffs, necessities, and paraphernalia onboard without my approval. That may sound like a trivial task, but I can assure you it is not. Just ask my fellow crewmates. Without me in command, we would have physical chaos onboard. Once we land, I continue to sustain my housekeeping role, making sure we maintain our visits in an orderly manner. No one departs from my plans without my permission. I also lead our team in physical development and body maintenance."

The colonists were now more comfortable with this wonderful, peace-loving band of aliens. *But we are more alien to this part of the cosmos than they are*, thought Kevin. Their trust was building, and they felt more at ease asking the myriad of questions they were bursting to probe.

"May we ask our questions at this point?" said Kevin.

"Okay," declared Okra. "Ask away."

"Why are you here?" asked Mike, a bit timidly.

"I believe I told you, but I will say it again," answered Orka. "First, to ensure you are here peacefully and are not interested in any hostile intentions, such as conquering any portion of the cosmos. Second, to aid you in assimilation into your new home."

"What is your interest in the six of us?" asked Jason.

"We want to welcome you to your chosen home, here in the Alpha Centauri solar system," answered Okra. "We are a peace-loving culture who travel the nearby stars and planets in our galaxy to ensure that our region is occupied with harmonious cultures."

"What if you determined we were hostile, rather than peaceful?"

"We would have not allowed you into our neighborhood."

"You would have prevented us from landing?"

"We would have stopped you from entering Alpha Centauri altogether as well as nearby star systems."

"Wow. You would have turned our spaceship around?"

"We would have destroyed your spaceship."

Suddenly the crew's fears went back up.

"You tracked us with two spacecraft," said Kevin. "Where is the other one?"

"They went on to visit Proxima Centauri b to investigate whether it is inhabited as yet."

"That was originally our destination until we were warned that it is being bombarded by solar flares from Proxima Centauri, eroding its atmosphere."

"Your warning was accurate. That planet may still be habitable. But this planet is much superior for you. Both the magnetosphere and the atmosphere will protect you well from any flares from Proxima Centauri."

"Have you ever visited Earth?"

"Several times."

"When?"

"The first times were hundreds of Earth years ago when your planet had a smaller population."

"Really. How old are you?"

"Each of us have lived hundreds of your Earth years."

"Wow. But you look so young. So, when did you first visit Earth?"

"The Old Testament of your Jewish and Christian Bibles document some of the early visits of our ancestors."

"Really?" said Michael. "I don't remember reading about any episodes of your visitations."

"You need to read it more carefully with an open mind. In particular, I refer you to Chapter Six of Genesis and much of Ezekiel.

"I'm sure you are aware of Stonehenge, the Great Pyramids and Sacsayhuaman in Peru?"

"Yes. Did you have something to do with their construction?"

"Our ancestors designed and orchestrated their erection to amuse themselves and to amaze the humans of that time. Although human evolution and technology have improved since then, even today humans are unable to duplicate these achievements. But they will eventually."

"When was the last time you visited Earth?"

"We performed surveillance of Earth several times during the past two hundred Earth-years. We even intended to land and visit with your culture. That is until we learned of what a hostile culture Earthlings had become. We observed your wars and inhumanity to each other. So many of your species are hostile to those that differ from them. It would appear hate, rather than love, dominates your cultures. We observed your nuclear weapons in silos in a state of readiness for the next wars. Eventually, humans will destroy each other, and the planet won't survive. We have lost all interest in visiting Earth. The human race and its flawed leaders are on a path of social, environmental, and physical annihilation."

"Yes. In part that is why we want to begin a new colony on this planet – to rid ourselves of the craziness of current life on Earth," stated Candice.

"We recognized your objective, which is why we support your endeavor and did not inhibit you from coming to this part of the galaxy."

"How do you live so long?" asked Jennifer.

"A big part of it has to do with evolution. Your Darwin called it survival of the fittest," answered Perna. "Many millennia ago, the status of our evolution was not that different from Earthlings today. Our ancient ancestors did not love their neighbors, but rather hated people different from themselves. This led to anxiety and strife between each facet of our civilization. Jealousy between the 'haves' and the 'have-nots' caused a civilization that strove to eliminate each other instead of working together for the common good. As you might surmise, this led to wars and more hatred. Eventually evolution took place. Both artificial selection and natural selection eliminated these conflicting beings, rendering them extinct. Only those who learned to live in love and harmony survived and thrived. This evolution has produced the utopia we enjoy today. This has been a significant contribution to our longevity.

"Another important contribution has been our lifestyle. We take diligent care of our bodies and what we put into them."

"What do you eat?" asked Samantha.

"Vegetables, leafy plants and some fruit."

"No meat or fish?"

"No. We dispensed with meat and fish from our diet when we learned such proteins produced enzymes that caused problems with our brains and development of anomalies in our organ tissue."

"What about dairy products?" asked Jennifer.

"Similar answer. We learned dairy products were the top source of saturated fat, contributing to heart disease, Type 2 diabetes, and Alzheimer's disease."

"What about nutrition for babies?" asked Sam.

"After breast milk, we learned to create better baby formulas than milk from any animal for maximum, healthy nutrition."

"So, you are all vegetarians, vegans in our terminology?"

"Yes, indeed."

"Is that why your skin color is green?" asked Jennifer.

"Probably so."

"All our animals were happy when we learned to give up meat, fish, and dairy products," added Perna, smiling.

"We were considering fishing and even hunting for food. Is that a bad idea?" asked Kevin.

"We would not prevent you from doing so. But you have a better alternative," suggested Orka.

"Stick to plants and fruits?"

"Yes."

"Why?"

"For health and longevity. I'll bet you had fewer health issues onboard your spacecraft when your diet was limited."

"Yes, you are right. If we continue to be vegans, will we live longer?" asked Jennifer.

"Most assuredly," said Perma. "But it will take millennia of evolution before you live as long as we. Actually, your successful journey proves you have a great start for evolutionary longevity. Your mutual tenets and commonality enabled solidarity and minimal anxieties between crewmates over the past eight years. You obviously love and respect each other. And your adoption of veganism will be critical. These have enabled a quantum leap in your evolutionary path."

"Are your bodies like ours?" Asked Jennifer. "Same organs? Same digestive system? Same circular system? Same respiratory system? Same neurological system? Same DNA?"

"Well, almost," responded Perma. "Our DNA is similar as are our organs. Our lungs and respiratory system differ. We breathe methane rather than air. But our organs and body systems are stronger, thanks to veganism."

"What kind of animals do you have on TRAPPIST-1e?" asked Candice.

"Countless kinds, perhaps thousands of varieties," answered Numbe. "Some are adopted as pets."

"Dogs and cats?"

"Oh, yes. They are quite common."

"Do they have long life spans too?"

"Well, not as long as ours, but longer than those on Earth."

"What do you do for fun?" asked Candice.

"Perhaps surprisingly, our entertainment is not that different from yours," said Twinto.

"We enjoy music, dancing, singing, and board games." Samantha listened intently, warming to these people.

"And of course, sex," admitted Numbe, shyly.

"Really? All these years?" asked Candice.

"Definitely," said Solpe enthusiastically.

"Are you monogamous?"

"We are now," answered Numbe. "We learned it eliminates a major cause of conflicts."

"Are your techniques any different from ours?" asked Candice, enthusiastically.

"Well, maybe a little. But now you are getting too personal."

"Let's get back to questions that may help you to assimilate," said Orka.

"Is the fresh water supply on Dreamworld scarce?" asked Samantha.

"Not at all," said Solpe. "Your cistern reservoir may have provided all you need but building the aqueduct from the river is an innovative idea. We are impressed. If Mike can turn that into an indoor plumbing system, you can all be impressed."

"We have only seen animals that are small with no fur and with smooth green skin. Why?" asked Kevin.

"Again, the answer is evolution. They don't need fur so long as they stay in this band of latitude where the temperature is moderate," answered Ampa. "If you venture closer to the poles, you will find animals with fur. The animals at this latitude have thick skin which protects them from the warmer summers and colder winters, but the dense atmosphere moderate temperature changes on this planet.

"The animals are small and green because of their diet. They only eat leafy green plants. If you decide to hunt them for food, you will learn of their soft, muscular meat. But you may be disappointed in the taste. It will not be like the steaks you consume on Earth."

"Aren't any animals carnivorous?"

"Not anymore. They too have learned to happily coexist and live off vegetables and fruits."

"Our rover will only take us sixty miles on a full charge. What can we see if we want to explore more of the planet?" asked Kevin.

Extraordinary things," answered Orka enthusiastically. "Although your colony is in an ideal temperate location, you must see the canyons beyond those hills. And the mountains and valleys beyond those. And the other oceans and beaches in the other hemispheres. And the polar caps."

"But how can we possibly get there?" asked Jason.

"You consider yourselves pioneers, don't you? Your ancestors found a way. In time, I'm sure you will as well."

"What about natural resources? Can we find minerals and materials to build the tools to create a better transportation vehicle?" asked Michael.

"Yes, but your rover can extend its range just by mounting your solar panels onboard," answered Ampa.

"Can we obtain solar power from Alpha Centauri A and Alpha Centauri B, not just from Proxima Centauri?" asked Kevin.

"Yes. Although both are approximately 1.2 billion miles from here, the luminosity of Alpha Centauri A is 50% brighter than your Sun, and the luminosity of Alpha Centauri B is only slightly less than your Sun. So, your solar panels can obtain more energy from these than you were able to obtain from your Sun when you were in the vicinity of Saturn. Of course, your primary energy source still comes from Proxima Centauri. Although it is a red dwarf with low luminosity, it is less than one hundred million miles from here. So, it remains your primary source of energy during daylight. At night, depending on your orbit, you will still be able to obtain some energy from Alpha Centauri AB."

"Need we be concerned about Proxima Centauri's solar flares?"

"Not really," said Orka. "That is why you were fortunate to select this planet. It has a strong magnetic field created by its molten core and rate of rotation. This field, in turn, creates a magnetosphere which protects the planet well from these flares, which can be quite strong. This planet's rich atmosphere also reduces the effects of these flares."

"Can you tell us about your home solar system?" asked Jason.

"I'll be happy to," said Orka. "TRAPPIST-1 is also a red dwarf. Its mass is only 9% of your Sun but it is much older: 7.6 billion years. We have seven planets in LaPlace resonance in its orbit, three of which are inhabited. Their orbital periods have precise numerical ratios: 8:5, 5:3, 3:2. 3:2, 4:3, and 3:2. All are tidally locked to our star. Although four planets are in the habitable zone and have water, only three have an atmosphere. These are the three that are inhabited. We live on TRAPPIST-1e. We enjoy the perpetual daylight and the tropical warmth generated by the tidal heating. Our atmosphere is sufficient to protect us from TRAPPIST-1 solar flares. Oceans cover less than half our planet. We have visited the other inhabited planets, TRAPPIST-1 d and f. But we prefer our TRAPPIST-1e. None of our planets have a moon."

"Is your planet anything like Dreamworld?" asked Samantha.

"It was more like Dreamworld many millennia ago," said Twinto. "Evolution has matured our environment: plants, animals, and of course, us into long-living, thriving entities. Also, like Dreamworld as you call it, our planet is rich in natural resources. Our atmosphere is almost 100% helium, argon, neon, ammonia, methane, and carbon dioxide with a trace of oxygen. Again, except for the atmosphere it is not that different from this planet. But we enjoy a magnetosphere even stronger than this planet."

"Do you plan to occupy Dreamworld in the future?"

"No. The abundance of oxygen in the atmosphere is toxic to us."

"Is that why you have those tubes in your nostrils?"

"Yes. The oxygen in this atmosphere would be lethal."

"Have you visited other inhabited planets?"

"Yes," said Orka. "Again, for your benefit I'll use the names your astronomers use so you may relate to them. First, one planet that is too far for us to visit right now, but much like your Earth is Kepler-442b. It is twice the size of Earth, but half the distance from its star as Earth. It orbits its star, a hot orange dwarf, Kepler-442 in just 112 days. But it is 1,200 light years away. Nevertheless, it is a high priority to visit sometime in the future for our decedents. We also intend to explore Kepler-186f, which is five hundred light years away. It is similar to your Earth, although its star is a red dwarf. We are reasonably sure it is home to intelligent life.

"The planets we have visited include Gliese 581g, 667Cc, Kepler-22b, Kepler-1649c and multiple planets of Wolf 1061. Each has a life form in some state of evolution. There are several more in the habitable zone less than 15 light years from Earth that are on our list to explore, although we have yet to confirm whether they are really inhabited. We would also like to visit LP 890-9c. It is the innermost planet, somewhat larger than Earth, in the habitable zone of an ultracool dwarf star. But it is one hundred light years away.

"We have also visited several large planets. You call them super Earths. One with life forms is Giese 581c. It is 5.5 times the mass of Earth, but only twenty light years away. Its life forms are just evolving so it does not yet have intelligent life. Another is LJS 1140b. It is 6.48 times the mass of Earth and lies forty-nine light years from here. It too has not yet evolved intelligent life. But it orbits extremely close to its star, a cool red dwarf. It has oceans of liquid water and a thick atmosphere protecting it from the radiation from its star. We recently visited a brown dwarf that orbits two small red giants. You call it VHS 1256 b. It is seventy-two light years from Earth and has water. But It does not have intelligent life, because its thick, turbulent atmosphere contains silicon sand."

"How can you tolerate such long durations to these planets?"

"Actually, we enjoy it. We all volunteered for these missions because we love to travel to new worlds and explore them. Ensuring the inhabitants are peace-loving is, of course, our primary objective. We were disappointed when we learned humans on Earth were war-like."

"Could we also visit other planets?" asked Jason.

"Yes," said Solpe. "But first you need to replenish your fuel for launches and landings. That should not be a problem since the regolith here is excellent for manufacturing your fuel. Also, until your longevity improves, I recommend limiting your visits to nearby planets. The three other planets of the Proxima Centauri system should be interesting and not difficult to reach. You might also consider the planets of Alpha Centauri A and Alpha Centauri B."

"But we were told these two stars have no planets."

"Your information is not at all accurate. Both stars have interesting planetary systems almost as rich as your

Sun. But we have yet to detect life on any of them. Still, you could explore them in reasonable timeframes."

"Are all the inhabitants on your planet like you?" asked Candice.

"Yes," replied Twinto. "That is why we live in harmony."

"What kind of government do you have?" asked Kevin.

"It is actually quite simple," answered Okra. "Our primary government is global. It is a democracy. It is not anything like your United Nations. Our interests are homogeneous without competing factions. The primary role of our global government is defense. Our expeditionary missions are part of that defense system. We are always mindful of any hostile worlds which do not share our peaceful, live-and-let live culture. Our sophisticated global defense system ensures no foreign entity can disrupt our harmony. My crew is part of this system. We provide detection and early warning of external threats. If we are not able to prevent aliens from disrupting our peaceful culture, we alert and/or dispatch our military. So, you see we play a vital role in maintaining our global society."

"How do you deal with crime and internal friction among outliers in your communities?" asked Jennifer.

"We really don't have crime and disorder between our residents," answered Perma.

"How is that possible?" asked Jason.

"You must understand how different our culture is from Earth's," said Numbe. "We raise our children to respect everyone. We teach them that we do not tolerate mutual disrespect, that relationships with their fellow beings are paramount. What you call love is expected with all their fellow beings."

"But what about the differences in your respective abilities? Doesn't that cause undesirable competitiveness and jealousy between your people?"

"Oh no. Not at all. Our children learn to respect and cherish their differences. They learn to appreciate the variety between their talents and are happy that the community benefits from their diverse talents."

"What is the population of TRAPPIST-1e?" asked Jennifer.

"Although our planet is slightly smaller than Earth, our population is slightly larger: more than eight billion inhabitants. It would be much larger if the ancient wars and revolutions had not decimated our ancient population. But our food source continues to grow. Also, our oceans are smaller than Earth's. Our land mass is 70% of the surface."

"Please tell us more about your children," asked Samantha.

"We have several generations of offspring," responded Perma." All are healthy. We visit them every chance we get. Also, we do video conferences which we have constructed over the millennia."

"Do you have political systems and organizations?" asked Kevin.

"Yes, but it is different than you have," said Orka. "First, we do not have political parties. Virtually all objectives and goals are well known and accepted at each level of government. We do have multiple levels: global, national, and local. Anyone can run for leadership positions. We have a pure democracy; each adult casts one vote for each elected office. This may sound unwieldy to you, but our communication system is quite advanced, giving each candidate a five-minute video of their qualifications sent to every voter. The votes are tabulated instantly and accurately. But since each candidate has almost identical objectives and goals, we simply vote for the most qualified candidate. Global leaders serve a ten-year term; national leaders for six years; local leaders for three years.

"As I mentioned earlier, we do have a global defense system to protect against any hostile aliens. But our nations are harmonious and do not require separate national departments of defense like Earth. Similarly, our three inhabited planets are quite compatible and have no conflicts. We all have the same objectives and goals. There is no reason for conflict. Our nations do have a variety of natural resources; so, we do trade with each other. Our local governments have a minimal infrastructure to educate the population on its laws, but with a harmonious population, there is no crime and there are no jails.

"Our citizens work to help each other, not to exploit individual advantages. Each citizen knows they have to do their best for the good of the community. Each of us has talents to contribute and is taught to do so to the best of our ability. Sounds like a utopia? Yes, it is, and it works."

"Wow!" said Kevin. "When the United States of America was founded, we might have been close to that kind of society. It is so pitiful to see what the USA has become. We can only hope that starting over in this new world will enable us to form a colony with similar attributes."

"It can be done," said Okra. "We wish you well."

"Do you have a monetary system?" asked Candice.

"No," said Okra. "We abandoned it because it was causing problems. Instead, we have a barter system. It works much better."

"How fast can your spacecraft go?" inquired Samantha.

"We can accelerate to 95% the speed of light almost instantly. But the G forces would be intolerable, even for us. So, we throttle back to a more reasonable level of acceleration, usually three hundred feet per second squared."

"But that's almost 10 Gs," said Mike. "We would pass out and maybe die at that rate of acceleration."

"We slowly trained our bodies to withstand those forces and more," said Solpe.

"So, you can't reach the speed of light?"

"Of course not. The laws of physics apply everywhere in the cosmos."

"What kind of propulsion system do you have?" asked Michael.

"Like your ion engine, it is nuclear based, but quite different; much more efficient."

"Do you grow your food onboard?" asked Samantha.

"Yes, but I answered that previously," responded Numbe. Like yours it is hydroponic but uses additives that promote nutrition and growth. You and I can meet separately later. I'm sure I can provide advice from our experience that will benefit you."

"What about your water on board?"

"We collect hydrogen and oxygen atoms en route and create water via stoichiometry as we go," answered Solpe.

"Is sleep a necessary period in your day to survive?" asked Samantha."

"Yes," answered Numbe." "But our sleep cycle is longer than yours. We need 12 hours per day to replenish our biological makeup. In part this is a major element of our longevity."

"Interesting. We usually get seven-to-eight hours sleep per day. If we slept longer, would we live longer?"

"I'm sure of it."

"Do you believe in God?" asked Michael.

"Of course," answered Twinto. "God created the cosmos, all the galaxies, stars, planets, and life. Once we determined that the Universe was created in a Big Bang, we realized this was done by God. No other explanation is logical. It is not hard to extrapolate the Universe backward in time to the singularity that manifested in the Big Bang. Prior to the Big Bang, no matter existed. So, God created the Universe from nothing except energy, and maybe not even energy."

"Do you and your people worship God?" asked Jason.

"Yes, regularly. We know that our life here will end, but we will have eternal life with God afterward. Worship enables us to communicate with God during our lives."

"Do you have a Savior?"

"Yes. He came down to our planet as one of us. Then he died to save us from our sins and rose from the dead, as will all of us."

"That sounds just like our Christianity."

"Of course. It is the same throughout the cosmos."

"Do you believe in the Holy Spirit?"

"Not in the sense that Christians do."

"Have you ever seen God?" asked Samantha.

"Unfortunately, no one currently alive has seen Him."

"Then how do you know that He exists?" asked Candice.

"We all have an internal truth-finder of sorts. We receive an inner calmness when we believe."

"Do you have any unbelievers?" asked Jason.

"Yes, especially among our young. But most ultimately acquire their beliefs as they mature."

"What advice do you have for us at this point?" asked Kevin.

"Continue to do much as you have been doing," answered Orka. "Nurture your love and mutual respect. Build your new home and respect nature. Do not allow any of your offspring to deviate from these tenets. Make sure all your colonists share these doctrines.

"IF there are no other questions, I suggest we pair up with our counterparts to delve further into how we can help you thrive in your new environment. I will meet with Kevin; Perma with Jennifer; Ampa with Mike; Numbe with Samantha; Solpe with Jason; Twinto with Candice."

"That sounds wonderful," said Kevin. "Thank you.

"How can we contact you if we need to follow up on anything?"

"All six of you should mentally concentrate on your desire for communication, and we may pick up the signal. Even if that doesn't work, we will periodically monitor your progress and make contact if we deem it necessary."

***

Two hours later, the "aliens" re-entered their spacecraft, slowly rose, then accelerated at an unbelievable speed into space.

"Wow!" exclaimed Kevin. "What did you think of that?"

"Unbelievable!" uttered Jennifer.

"How could we be so fortunate?" said Michael. "Having super intelligent friendly neighbors."

"And so helpful and constructive," added Samantha.

"And interesting," said Candice.

"I think they just gave us a lesson in how to live our lives," added Jason.

"I agree," said Kevin, "Well, it is time for another vote. Shall we adopt veganism?"

"I guess I could forgo my craving for steak," said Mike. "If we accept everything they said, the benefits out-weigh the indulgences."

"Well, they certainly know much more than we do," said Kevin. "In my opinion we'd be foolish to ignore their advice."

"So, time to vote."

Once again it was unanimous; all six agreed to accept veganism. Although Sam was concerned whether veganism would be best for her baby when she weaned her from breastfeeding. She hoped the tips she received from Numbe would suffice.

"I can't wait to let NASA know about our brilliant extraterrestrial creatures and what they have taught us," said Kevin.

"Let's wait awhile and digest this stuff," suggested Mike. "This is so important that we need to be careful in informing Earth."

Kevin wondered what Mike meant but was amenable to the request. It would take over four years for Earth to receive the information, so there was no reason to rush.

Mike went back to finishing the pipeline with Homer. Samantha marveled at his skill. In less than one week he finished the pipeline and its rudimentary pump, and Mike began constructing the internal plumbing in each igloo.

Kevin and Jason mounted the solar panels onto the rover, enabling virtuously limitless range in daylight.

A few days later Mike recruited Jennifer and Candice. Samantha was ready to give birth. Thanks to the exercises prescribed by Candice, the birth was reasonably easy. Sam and Mike gave birth to a healthy seven-pound baby girl. Sam's fears of defects from space travel were dispelled when she saw and held her beautiful baby. Mike was so proud, no one could wipe the smile from his face. The first citizen of Dreamland was welcomed into the colony. Michael baptized her Aurora Nova Stone.

Soon afterward, both Jennifer and Candice announced that they too were pregnant. The colony's procreation had begun. "We can't wait for all of us to be teachers for all our children without the crazy wok pressure, gender diversity, and Critical Race Theory messing up their malleable brains," said Jennifer.

But Kevin reminded the team of their unfinished business: reporting back to NASA of their unbelievable success in finding a planet so conducive for a flourishing colony. They met in Kevin and Jen's igloo and sat around the table, Sam holding Aurora.

"It's time to announce our landing and safety back to NASA," said Kevin. "They have to be worried since they think we proceeded on to Alpha Centauri b but don't even know whether we landed safely."

"Wait a minute," Kev, said Mike. "Why is that necessary? We have a wonderful thing going here and we don't want to mess that up."

"What do you mean?" asked Kevin. "How would that mess anything up?"

"NASA is bound to augment our colony with another spacecraft and crew, maybe many of them, once they learn what a perfect planet this is," said Jason.

"And with all the terrible conditions on Earth right now, they will have no trouble recruiting," added Candice.

"Okay, so what's wrong with that?"

"Kev, we successfully survived this mission because of our compatibility, overcoming human frailties like ideological differences, competitiveness, jealousy, and religious diversity," said Mike. "Remember how hard we worked to ensure harmony when we signed on? We probably could not have succeeded without this close-knit camaraderie."

"But we owe it to NASA, and the nation for that matter, to notify them that their grand plan was successful. If I don't at least send them a pulse from our laser, they will never know where we are or our amazing success."

"I agree with Kevin," said Jennifer. "It is a moral obligation that we can't shirk. They recruited us, supported us, and financed us. We can't reward them with the selfish act of keeping this Shangri-La to ourselves."

"Kevin, think through what would happen if NASA sent one or more missions to join us," said Mike. "Do you think they would select people based on compatibility with us, or acquiesce to America's pressure for diversity? Remember the criticism NASA received because we were so alike in pollical philosophy?

"If we don't send any signal or message back to Earth, NASA will believe we landed on Alpha Proxima b and died from radiation poisoning. If so, they will never send another spacecraft our way."

"I agree with Mike," said Candice. "Instead of sending people who would assimilate into our thriving colony, they would have a diversified agenda at odds with our own. We will not have any input into the next crew selection. It would create a US versus THEM society, much like America has today. We came here in part to escape this divisive conflict. We can't allow its recreation on Dreamland. We can take a lesson from our alien friends about social conflict. I want to bring our children into this world in harmony, not Earth-like conflict."

"What do you think, Sam?" asked Mike.

"I'm conflicted," Sam said. "I do feel an obligation to both America and NASA. We owe them feedback on the success of their mission. Moreover, they alone gave us this opportunity. We certainly would not be here if it were not for them. Yet Mike and Candice make powerful points on assimilation versus conflict between different ideologies. I don't want my daughter to experience the unhealthy conflicts taking place in America today."

Jason strongly agreed. "We need to listen to our alien friends. We have made a quantum leap in positive evolution of our species. We can't screw it up by backtracking with diversity."

"Well, we clearly have a dilemma, gang," declared Kevin. "I feel a major allegiance to NASA, but I also feel a big obligation to all of you. I would like us to be of one mind in this crucial decision, just as we have in all the decisions we made on our journey. Let's sleep on it and continue our discussion tomorrow."

Kevin continued to withhold the secret that the primary reason for the mission was to ensure the perpetuation of the human race in the event of Armageddon on Earth. He didn't think disclosing this additional objective would change anyone's mind. In fact, it could solidify the chasm.

Kevin did not sleep well that night. Nor did the rest of the colony. Their future was at stake and obviously all six were stakeholders. This decision would not be resolved overnight. He was sure if he asked the aliens for guidance, they would strongly recommend against giving Earth information that might compromise planet harmony with diversity. They might even prevent a contrarian crew from entering their domain. Kevin's stomach was turning. He prayed for the wisdom of Solomon.

*****

Acknowledgements

Once again, I am pleased to acknowledge the excellent job of proofreading the manuscript by Peggy Williams. I am also grateful to my wife for her patience and suggestions. Finally, I am grateful for the ideas, discussions, and encouragement from several residents of the Reata Glen community in Rancho Mission Viejo, California.

My granddaughter, Taylor Odish, designed the cover. She is a professional graphic designer who also created the covers of my previous books.

Wikipedia has been this book's benefactor with its wealth of information to ensure factual accuracy throughout.

Bibliography

Krause, Kurth (2019), My 36 Years in Space: An Astronautical Engineer's Journey through the Triumphs and Tragedies of America's Space Programs, Costa Mesa, California

Odenwald, Sten (2022), Interstellar Travel: An Astronomer's Guide, The Astronomy Café

Ross, Hugh N. (1998), The Genesis Question: Scientific Advances and the Accuracy of Genesis, Colorado Springs, Colorado, Nav Press

Strobel, Lee P. (1998), The Case for Christ, Grand Rapids, Michigan, Zondervan

About the Author

This is the first novel by Kurth Krause. His previous books are non-fiction. His first book, *My 36 Years in Space,* documents Krause's academic and career background which were fundamental in describing the technology required for authoring this book. *My 36 Years in Space* received unanimous five-star reviews on Amazon. The next page presents the formal review of his first book by a professional reviewer.

Kurth Krause was born in Milwaukee in 1940. He received his Bachelor of Science degree in mathematics and physics from the University of Wisconsin in 1962. He studied astronautical guidance in graduate school at the Massachusetts Institute of Technology in 1965. Kurth received management training at the UCLA Graduate School of Management and the Stanford Executive Institute. He has been married to Susan (Firle) Krause since 1963. They reside in Rancho Mission Viejo, California. They have two children and four grandchildren.

Krause became a pioneer in the space field; at age 25, while at MIT he wrote vital subroutines of the guidance and navigation software for the Apollo Command Module and Lunar Module flight computers which flew in every Apollo mission. He and Sue witnessed the launch of Apollo 11 from the astronaut viewing area at Cape Canaveral. At age 28, Krause sat at the guidance console in the Houston backup Mission Control Room while Neil Armstrong and Buzz Aldrin landed on the Moon.

Official Review: *My 36 Years in Space* by Kurth Krause

The following is an official OnlineBookClub.org review by InStoree. » 4 Dec 2019

10 … 9 ... 8 ... Engine Ignition! 7 ...6 ... 5... 4 ... 3 ... 2 ... We have liftoff!

This countdown would launch Apollo 11 toward a mission the media called "the greatest accomplishment of the decade." On July 20, 1969, man first stepped on the Moon, a historical event watched around the world. Through the lens of Kurth Krause's space engineering career, readers come to know how events unfolded during one of the most dangerous space missions ever attempted, as highlighted in *My 36 Years in Space: An Astronautical Engineer's Journey through the Triumphs and Tragedies of America's Space Programs*. A team of masterminds managed to land two astronauts on the Moon's surface. But would they be able to bring them back safely?

Launching a rocket into space is a complex process, but it is not as complicated as guiding it back to Earth from another celestial body in our Solar System, such as the Moon. As a young engineer, Kurth Krause is tasked to develop an onboard program to be used on the Command Module (CM) and the Lunar Module (LM) to facilitate a return-to-Earth mission during the Apollo Era. It all begins in a rather technologically primitive period when computer science, missile engineering curricula, and space science were still limited. Krause highlights some of the challenges and space achievements that America went through at a time when the space race was a determinant for global dominance and supremacy. In addition, the writer gives a glimpse into his personal family life and how he balanced this with his highly demanding work.

This book was an exhilarating read for me, making me aware of mind-blowing scientific principles. The absorbing storytelling kept me completely engaged. With my minimal knowledge of astronomy and space programs, I was excited to better understand the astronautical wisdom that the author poured into his work, complements of a thirty-six-year career in the space field. Reading about the intellectual accomplishments of these men was truly awe-inspiring. Delving into this non-fiction piece may seem intimidating, even burdensome. The text contains scientific terms that were a bit complicated but not incomprehensible.

If you are not a "rocket science" geek but wish to better understand the space world, this book provides a thrilling and comprehensible journey through America's space programs. I managed to navigate the complexity of this work with little difficulty. Somehow it became easier to understand, and in the end, a feeling of absolute marvel took over me. It was impressive.

Krause thoroughly describes the difficult mission and the amazing triumph of Armstrong, Buzz Aldrin, and the entire space crew. I imagined how glorious this event must have been — especially viewing it live in 1969. I thought of Neil Armstrong's first words while stepping down on lunar soil: [sic] "One small step for a man, one giant leap for mankind." The author self-describes in the book as someone whose "doodling included playing with Fibonacci numbers, infinite series, singularities, and deriving trig identities." Anyone who relates to that will definitely enjoy the book. But you don't need a background in math or science to find the work compelling and understandable. It can also be recommended to students who aspire to a career in space program engineering; it will give you a closer look at the Apollo Space Program, the NASA customers, and an aerospace career.

I encountered no grammatical errors, and the well-edited book allowed for easy reading. The absence of expletives was no surprise for a book of this class. The author also made good use of pictures, which made it easier for me to grasp the scientific concepts being described. *My 36 Years in Space*, by Kurth Krause, deeply deserves a 4 out of 4 stars rating to the Moon and back.

www.ingramcontent.com/pod-product-compliance
Lightning Source LLC
Chambersburg PA
CBHW020525310726
48979CB00014B/2221/J

* 9 7 8 0 9 9 8 4 5 6 8 6 7 *